CLOSTRIDIUM DIFFICILE - A COMPREHENSIVE OVERVIEW

Edited by **Shymaa Enany**

***Clostridium Difficile* - A Comprehensive Overview**
http://dx.doi.org/10.5772/65578
Edited by Shymaa Enany

Contributors

Luminița Smaranda Iancu, Andrei Cârlan, Ramona Gabriela Ursu, Maria Tomas, Laura Fernandez-García, Lucia Blasco, Maria Lopez, Alex Therien, Mónica Oleastro, Joana Isidro, Aristides Mendes, Mónica Serrano, Adriano Henriques, William Sherman, Christopher Lewis, Jong Lee, David Herndon, Shymaa Enany

Published by InTech
Janeza Trdine 9, 51000 Rijeka, Croatia

Publishing Process Manager Mirena Calmic
Technical Editor SPi Global
Cover InTech Design team

First published September, 2017
Printed in Croatia
Legal deposit, Croatia: National and University Library in Zagreb

Additional hard copies can be obtained from orders@intechopen.com

Clostridium Difficile - A Comprehensive Overview, Edited by Shymaa Enany
p. cm.
Print ISBN 978-953-51-3427-5
Online ISBN 978-953-51-3428-2

For EU product safety concerns:
IN TECH d.o.o., Prolaz Marije Krucifikse Kozulić 3, 51000 Rijeka, Croatia,
info@intechopen.com or visit our website at intechopen.com.

Meet the editor

Dr. Shymaa Enany is an assistant professor of Microbiology and Immunology at the Suez Canal University, Egypt. She received her PhD degree from the School of Medical and Dental Sciences, Niigata University, Japan, and completed her postdoctoral work in collaboration with many laboratories in San Diego, California, USA, and in Niigata, Japan. She is an editorial board member and a reviewer in many journals and scientific associations and has many publications in eminent journals as well as books. She has as an excellent experience in bacterial genomics and proteomics.

Contents

Preface

Clostridium difficile , or you can call it as *C. difficile* or *C. diff,* is a bacterium living everywhere around us, in the air, water, and soil and in the feces of humans and animals, living normally together with other microorganisms inside human alimentary canal in a balance number to each other. People can easily become infected with *C. difficile* if they touch contaminated clothing, sheets, or other objects and then touch their mouths. Also, when people take antibiotics to knock out a certain bacterial infection, these antibiotics can trigger a disturbance in the bacterial balance inside their gut leading to overgrowth of *C. difficile* causing symptoms ranging from diarrhea to life-threatening inflammation of the colon.

C. difficile infections are most often spread in the healthcare facilities between workers, patients, and residents.

Each year in the United States, almost a half million people get sick from *C. difficile,* and approximately 29,000 patients died within 30 days of its initial diagnosis. Nowadays, *C. difficile* infections have become more frequent, severe, and difficult to treat.

Therefore, we decided to write this book to discuss the numerous diagnosis methods and the suitable treatment presented by international leaders in their respective fields.

This book consists of six review chapters. Each chapter starts with a brief introduction, including its aim, and then goes on to provide detailed information about current research relevant to the field. The first chapter is the introductory one that gives an overview of the *C. difficile* as an important pathogen to allow the reader to form a complete picture about this bacterium and its subsequent infection. Through the chapters within, the authors explored *C. difficile* life cycle including growth, spore formation, and germination. They examined *C. difficile* epidemiology and the different antimicrobial resistance patterns. Recent developments in treatment and prevention of *C. difficile* are also reviewed here.

We believe that our book is an excellent one for microbiologists, especially those who are interested in *C. difficile.* We hope you enjoy reading it. Finally, we would like to thank all the contributing authors without whose dedication and brilliant research, this project would not have been accomplished.

Dr. Shymaa Enany
Department of Microbiology and Immunology, Faculty of Pharmacy
Suez Canal University
Ismailia, Egypt

Chapter 1

Introductory Chapter: *Clostridium difficile* Infection Overview

Shymaa Enany

Additional information is available at the end of the chapter

http://dx.doi.org/10.5772/intechopen.69983

1. Introduction

Clostridium difficile (*C. difficile*) is an anaerobic, spore-forming Gram-positive bacillus that was first described in 1935 as part of the intestinal microflora in neonates, even so it was not identified as a causative agent of human disease until 1978 [1]. The clinical presentation of *C. difficile* infection (CDI) could be asymptomatic, mild or moderate diarrhea and fulminant colitis [2, 3].

2. Epidemiology of CDI

Center of Disease Control and Prevention showed an elevation in the incidence and the severity of CDI [4]. More than 250,000 person need to be hospitalized due to CDI, and around 14,000 people die from it in the United States every year [5]. Among hospitalized patients, the incidence of CDI differs every year and from location to another. It has been elevating, to nearly 15 per 1000 hospital dismissal [6] and around 20 cases per 100,000 individual in the community [7]. *C. difficile* can only colonize the gut when the normal intestinal microflora is changed by the usage of antibiotics and that was proved by the 16S ribosomal RNA sequencing [8]. Therefore, the antibiotics usage remains the most important risk factors for *C. difficile* infection. Many antibiotics are associated with the CDI such as ampicillin, amoxicillin, cephalosporins, clindamycin, fluoroquinolones, trimethoprim and sulfonamides [9]. Another risk factor for CDI is the age; the severity of the infection increases as the age increases [10]. Poor hand hygiene has also previously been shown to play a part in CDI transmission [11]. Hospitalization considers also a main risk factor as it brings together many CDI risk factors in one place such as the use of antibiotics, the spore contaminated media, inappropriate hand hygiene and the elderly patients [12].

3. *C. difficile* virulence factors

C. difficile has many virulence factors including toxins, sporulation, surface layer proteins and adherence. It produces many toxins such as the enterotoxin TcdA, the cytotoxin TcdB and the binary toxin CDT [12]. Theses toxins cause disruption of the actin cytoskeleton and tight junction and cause a decrease in the transepithelial resistance, fluid accumulation and damage of the intestinal epithelium [13].

4. Diagnosis of CDI

Diagnosis of *C. difficile* is easily done in the laboratory and usually performed for the patients suffering from diarrhea. Currently, CDI is diagnosed by several available diagnostic tests such as enzyme immunoassay (EIA), EIA for *C. difficile* glutamate dehydrogenase (GDH) or by DNA-based tests which recognize the genes of *C. difficile* toxin in the stool sample. Additional diagnostic tests are available like toxigenic cultures and cell culture neutralization assays [14].

Stool culture for *C. difficile* requires anaerobic culture and is not widely available [9]. Radiography suggestive of CDI includes polypoid mucosal thickening, haustral fold thickening or gaseous distention of the colon; however, radiographic features are not sensitive and not CDI specific [15]. Another diagnostic method is the endoscopy which is rarely required, but it may be helpful in case of doubt of CDI from the clinical signs with all the laboratory tests showed negative results or in patients with inflammatory bowel disease [16].

5. Prevention and treatment of CDI

Since there is no effective vaccine for the CDI control, prevention of the CDI has been a demand and it has focused on barrier methods and environmental hygiene in a trial for prohibiting *C. difficile* spores and for the reduction of the CDI risk factors: isolation of CDI patient in private room, gowns and gloves usage, hand hygiene and the use of sporicidal solution for rooms [12]. Moreover, altering the antibiotic prescribing could be a good way for preventing CDI spreading, since the possibilities of some antibiotics to stimuli CDI are smaller than others. Furthermore, the use of probiotics to prevent CDI could be a safe method. Probiotics usually formed of live microorganisms which give a lot of health benefits to the patient. These microorganisms work through direct activity against *C. difficile* through the inhibition of the bacterial adherence, the modification of the response of the host and the induction of production of specific IgA antitoxin [17, 18].

The treatment of CDI has not shown a big variation. For the acute infections, metronidazole and oral vancomycin has been the mainstay of treatment since 1970. Fidaxomicin was approved in 2011 by the Food and Drug Administration for CDI treatment [19]. Treatment of the first recurrent CDI infection is recommended with a repeat course of either metronidazole or vancomycin, and this regime is proved to be successful in 50% of patients [20]. Second recurrent infection can be treated with fidaxomicin which proved to prevent further episodes of *C. difficile* [21]. The fecal microbial transplantation is one of the bacterio-therapy used to prevent CDI. It is referring to the infusion of fecal suspension from a healthy person to reinstating the gut microbiota of the recipient.

6. Conclusion

Since the CDI causes common and serious problems, many researchers have focused on improving the prevention and the treatment of CDI. In this book, we have focused on studying the pathogenesis and the virulence factors of *C. difficile* including toxins and trying to explore the different diagnostic tools and preventive therapeutic methods.

Author details

Shymaa Enany

Address all correspondence to: shymaa21@yahoo.com

Microbiology and Immunology Department, Faculty of Pharmacy, Suez Canal University, Egypt

References

[1] Bartlett JG. *Clostridium difficile*: History of its role as an enteric pathogen and the current state of knowledge about the organism. Clinical Infectious Diseases : An Official Publication of the Infectious Diseases Society of America. 1994;**18**(Suppl 4):S265-S272

[2] Bartlett JG. Clinical practice. Antibiotic-associated diarrhea. The New England Journal of Medicine. 2002;**346**:334-339

[3] Kelly CP, LaMont JT. *Clostridium difficile*—More difficult than ever. The New England Journal of Medicine. 2008;**359**:1932-1940

[4] McDonald LC, Owings M, Jernigan DB. *Clostridium difficile* infection in patients discharged from US short-stay hospitals, 1996-2003. Emerging Infectious Diseases. 2006;**12**: 409-415

[5] Centers for Disease Control and Prevention. Antibiotic resistance threats in the United States; 2013. Accessed December 15, 2015

[6] Steiner C, Barrett M, Weiss A.HCUP Projections: *Clostridium difficile* Hospitalizations 2001 to 2013. Rockville, MD: Agency for Healthcare Research and Quality; 2014

[7] Chitnis AS, Holzbauer SM, Belflower RM, Winston LG, Bamberg WM, Lyons C, et al. Epidemiology of community-associated *Clostridium difficile* infection, 2009 through 2011. JAMA Internal Medicine. 2013;**173**:1359-1367

[8] Dethlefsen L, Huse S, Sogin ML, Relman DA. The pervasive effects of an antibiotic on the human gut microbiota, as revealed by deep 16S rRNA sequencing. PLoS Biology. 2008;**6**:e280

[9] Leffler DA, Lamont JT. *Clostridium difficile* infection. The New England Journal of Medicine. 2015;**373**:287-288

[10] Lessa FC, Gould CV, McDonald LC. Current status of *Clostridium difficile* infection epidemiology. Clinical Infectious Diseases : An Official Publication of the Infectious Diseases Society of America. 2012;**55**(Suppl 2):S65-S70

[11] Rutala WA, Weber DJ. Role of the hospital environment in disease transmission with a focus on *Clostridium difficile*. Healthcare Infection. 2013;**18**:18-22

[12] Rupnik M, Wilcox MH, Gerding DN. *Clostridium difficile* infection: New developments in epidemiology and pathogenesis. Nature Reviews Microbiology. 2009;**7**:526-536

[13] Rupnik MJ. The Comprehensive Sourcebook of Bacterial Protein Toxins. 3rd edn. Burlington, Massachusetts, USA: Academic Press; 2006

[14] Kufelnicka AM, Kirn TJ. Effective utilization of evolving methods for the laboratory diagnosis of *Clostridium difficile* infection. Clinical Infectious Diseases : An Official Publication of the Infectious Diseases Society of America. 2011;**52**:1451-1457

[15] Ofosu A. *Clostridium difficile* infection: A review of current and emerging therapies. Annals of Gastroenterology. 2016;**29**:147-154

[16] Johal SS, Hammond J, Solomon K, James PD, Mahida YR. *Clostridium difficile* associated diarrhoea in hospitalised patients: Onset in the community and hospital and role of flexible sigmoidoscopy. Gut. 2004;**53**:673-677

[17] Banerjee P, Merkel GJ, Bhunia AK. Lactobacillus delbrueckii ssp. bulgaricus B-30892 can inhibit cytotoxic effects and adhesion of pathogenic *Clostridium difficile* to Caco-2 cells. Gut Pathogens. 2009;**1**:8

[18] Qamar A, Aboudola S, Warny M, Michetti P, Pothoulakis C, LaMont JT, et al. Saccharomyces boulardii stimulates intestinal immunoglobulin A immune response to *Clostridium difficile* toxin A in mice. Infection and Immunity. 2001;**69**:2762-2765

[19] Leffler DA and Lamont JT. *Clostridium difficile* Infection. The New England journal of medicine. 2015;**373**:287-8

[20] Leffler DA, Lamont JT. Treatment of *Clostridium difficile*-associated disease. Gastroenterology. 2009;**136**:1899-1912

[21] Cornely OA, Miller MA, Louie TJ, Crook DW, Gorbach SL. Treatment of first recurrence of *Clostridium difficile* infection: Fidaxomicin versus vancomycin. Clinical Infectious Diseases : An Official Publication of the Infectious Diseases Society of America. 2012;**55** (Suppl 2):S154-S161

Overview of *Clostridium difficile* Infection: Life Cycle, Epidemiology, Antimicrobial Resistance and Treatment

Joana Isidro, Aristides L. Mendes, Mónica Serrano,
Adriano O. Henriques and Mónica Oleastro

Additional information is available at the end of the chapter

http://dx.doi.org/10.5772/intechopen.69053

Abstract

The use of antimicrobial agents and acquired resistances explains in part the emergence and spreading of epidemic strains of *Clostridium difficile*. Continued use of antimicrobial therapy still represents an acute danger in triggering the emergence and spreading of new resistant and multiresistant strains including against first-line antibiotics. We examine the pathway of peptidoglycan synthesis in this organism and associated resistances, as well as resistance to other classes of antibiotics. The life cycle of *C. difficile* involves growth, spore formation and germination. Spores endow the organism with a formidable capacity of persistence in the environment and in the host, resistance, dissemination and infectious potential. Highly resistant spores produced by antibiotic-resistant/multiresistant strains may be one of the most serious challenges we face in what concerns the containment of *C. difficile*. Finally, we review recent developments in the treatment and prevention of *C. difficile* infection.

Keywords: *Clostridium difficile* infection, antibiotics, epidemiology, drug resistance, spores, β-lactam antibiotics, fidaxomycin, *Clostridium scindens*

1. *Clostridium difficile* life cycle: antibiotic-resistant spores as infectious, persistence and dissemination vehicles

1.1. *Clostridium difficile* biology

The human gut is the home of a community of as many as 1000 species of commensal, beneficial and pathogenic microorganisms. Recent studies suggest that at least half of the bacterial genera found in the gut produce resilient spores [1]. One of these organisms is *Clostridium difficile*, a Gram-positive spore-forming anaerobe. *C. difficile* was recently placed in the Peptostreptococcaceae family and renamed *Peptoclostridium difficile* [2] and later *Clostridioides difficile* [3], yet we use herein the still more familiar designation of *C. difficile*.

Although *C. difficile* (*Bacillus difficile* by that time) has been first described in 1935 by Hall and O'Toole as part of the bacterial flora of the meconium and faeces of infants [4], it was only in 1974 that three independent studies have implicated *C. difficile* has an important cause of disease in humans. These studies also showed that the organism produced a toxin that was highly lethal to mice, but the toxicity to humans was not demonstrated. In one of these studies, the causal effect of antibiotic exposure and gut disease was demonstrated. The study, conducted by Tedesco and co-authors, found a significant association between patients receiving clindamycin and the development of pseudomembranous colitis (PMC), although at this stage the etiology of this observation remained unknown [5]. In 1977, *C. difficile* was isolated from the faeces of hamsters with clindamycin-induced colitis, confirming this pathogen as the cause of antibiotic-induced disease in animals [6]. In 1978, two studies provided confirmation of the causal association of *C. difficile* infection (CDI) and antibiotic-associated PMC in humans [7, 8].

The original name of *C. difficile* reflects the difficulties in the isolation and growth of this bacterium in the laboratory. Nowadays, we face the problem of being unable to control the growth and spreading of this organism. In the last decade, epidemic strains, including those of ribotype 027 (RT027), have emerged that caused outbreaks associated with increased disease severity and higher recurrence, morbidity and mortality rates, and *C. difficile* is now considered the major causative agent of nosocomial diseases associated to antibiotic therapy in adults [9–11]. *C. difficile* causes close to 500,000 infections and 29,000 deaths each year in the United States alone, with about 20% of CDI leading to recurrence, and imposing an economic burden on the healthcare system estimated at over US$ 1 billion [12, 13] or an estimated €3 billion in Europe [14]. *C. difficile* is categorized as an urgent threat, the highest level of concern, by the CDC (Center for Disease Control and Prevention). Moreover, increased rates of community-associated *C. difficile* disease, affecting groups not previously at risk, such as children, healthy young adults and pregnant women, and zoonotic transmission are a raising concern [11].

1.2. The life cycle

C. difficile is an enteric pathogen that relies on the disturbance of the normal gut microbiota to expand in the gut and cause infection; individuals with a normal, balanced microbiota are usually resistant to infection by *C. difficile* [14–16] (see below). Unlike most of the commensals, *C. difficile* resists to a wide range of antibiotics (see below). Resistance to antibiotics such as erythromycin, chloramphenicol or tetracycline is largely mediated by transposons that are present in the *C. difficile* genome [17–19]. Individuals with a normal, balanced microbiota are usually resistant to infection by *C. difficile* (see subsequent text). Disease symptoms range from mild diarrhoea and abdominal pain to life-threatening inflammatory lesions such as PMC, toxin megacolon or bowel perforation, and in severe cases sepsis and death [16, 20, 21]. These symptoms are mainly caused by two potent proinflammatory cytotoxins, TcdA and TcdB, that following release from the bacterium, translocate to the cytosol of target host cells and inactivate, by monoglucosylation, small GTP-binding proteins, including Rho, Rac and Cdc42. TcdA and TcdB cause actin condensation, disintegration of the cytoskeleton, cell rounding and eventually cell death [22]. These toxins are coded for by two genes, *tcdA* and *tcdB*, located in the pathogenicity locus (PaLoc) (**Figure 1A**). The PaLoc also contains three additional genes, *tcdE, tcdR* and *tcdC*. TcdE is thought to be a putative holin-like protein involved in toxin secretion; its impact on toxin secretion, however, is still under debate [23, 24]. TcdR is an RNA polymerase sigma factor that serves as the main positive regulator of expression of the PaLoc and activates its own expression from two tandem promoters [25, 26] (**Figure 1A**). Importantly, the expression of *tcdR* is also activated from a promoter responsive to σ^D, the main regulatory protein involved in the final stages of flagellar assembly [27] (see also below) (**Figure 1A**). TcdC is thought to be a TcdR-specific anti-sigma factor that negatively regulates TcdR-dependent transcription [23, 28, 29]. TcdC was also shown to bind DNA, which suggests an alternative function for this anti-sigma factor [30]. Some *C. difficile* strains, as those of RT027 and RT078, also produce a binary toxin known as CDT (*C. difficile* transferase) (**Figure 1B**). The CDTb component of CTD binds to the host cells and translocates CTDa, that ADP-ribosylates actin, inducing depolymerization of the actin cytoskeleton. Toxin-induced actin depolymerization also induces redistribution of microtubules and formation of long microtubule-based protrusions at the surface of the intestinal epithelia cells; these protrusions trap the bacteria in small compartments, increasing the adherence of *C. difficile* [31–33]. In mice, by inducing inflammation via a Toll-like receptor 2 (TLR2), CDT suppresses a protective colonic eosinophilic response [34].

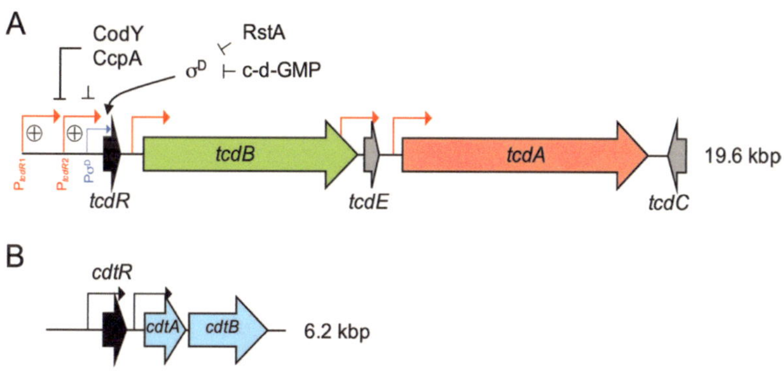

Figure 1. Schematic representation of the PaLoc (A) and CdtLoc (B) loci of *C. difficile*. A: This shows the genetic organization of the PaLoc in toxinogenic strains, which includes the *tcdR, tcdB, tcdE, tcdA* and *tcdC* genes. *tcdR* codes for an RNA polymerase sigma factor that controls expression of the *tcdB* and *tcdA* genes and possibly also of *tcdE*. The main promoters in the region are represented by broken arrows. The figure highlight the main regulatory circuits highlighted in the text. Transcription of the *tcdR* gene is governed by at least three promoters, two of which are auto-regulatory ('+' signs); a third promoter is under the control of σ^D, the regulatory protein governing the last stages in assembly of the flagellum. A promoter recognized by the housekeeping sigma factor σ^A, located downstream of the σ^D-type promoter has also been proposed [108]. Transcription of *tcdR* is directly repressed by CodY and CcpA but the latter also represses the expression of other PaLoc genes (not shown) [103, 108]. c-d-GMP also represses the expression of the PaLoc by inhibiting the production of σ^D. B: This shows the organization of the binary toxin locus or CdtLoc. *cdtR* codes for a response regulator required for full expression of the downstream genes, *cdtA* and *cdtB*, coding for the two components of the binary toxin CDT. The putative kinase that activates CdtR is unknown. The CDT toxin is only produced by some strains, including those of RT027 and RT078; in some strains, the locus is absent, whereas in others, such as 630Δerm, *cdtA* and *cdtB* are pseudogenized [33].

Since *C. difficile* is a strict anaerobe, its virulence potential is linked to the ability to form spores. Spores are resistant to heat, oxygen and other environmental insults, including commonly used ethanol-based disinfectants. When ingested, the spores are able to pass the gastric barrier and reach the intestine where they are thought to attach to the epithelial cells in order to achieve proper germination, which is induced by certain bile salts (see below) (**Figure 2**). Spore germination and outgrowth, in the absence of a competitive microbiota, will result in the establishment of a population of vegetative cells that will expand, produce the TcdA and TcdB toxins and eventually more spores (**Figure 2**). The toxins will cause damage to the colonic mucosa and eventually severe diarrhoea; shedding of the spores to the environment allows the infection of new hosts (**Figure 2**) [15, 22, 35, 36].

Spores are the vehicle for transmission as well as for environmental persistence. Mice exposed to spores exhibited recurrent infection with the same strain (disease relapse), but a *spo0A* mutant, lacking the key regulatory protein governing entry into sporulation and thus unable to form spores, was incapable of recolonization and host-host transmission [37]. Evidence suggests that *C. difficile* forms biofilms *in vivo* and *in vitro* and that the main virulence and persistence factors (toxins and spores) are produced within these structures [38–40]. In mice, infection with *C. difficile* spores followed by a 7-day period of treatment with clindamycin results in entry into a highly contagious period, during which high number of spores are shedded [40]. For a RT027 strain, the 'supershedder' state remained for months, even after

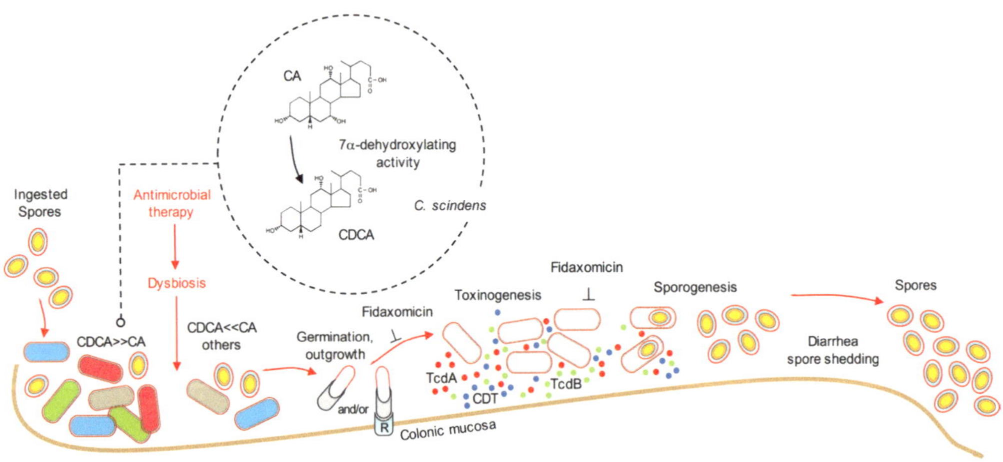

Figure 2. Schematic representation of the *C. difficile* infectious cycle. Spores are ingested and are able to pass the gastric barrier to reach the intestine. In the intestine, when the normal gut microbiota is disturbed, as during antibiotic treatment, the ratio of the bile salts derivatives cholate (CA) and chenodeoxycholate (CDCA) increases and the spores will germinate. *C. scindens*, for example, produces a 7α-dehydroxylating activity that converts CA into CDCA, blocking germination of *C. difficile* spores (insert). Spore germination, which may occur following recognition of a receptor (R) in the colon, and cell outgrowth will eventually produce a population of vegetative, actively growing cells, which will initiate toxinogenesis and sporogenesis. The two cytotoxins TcdA and TcdB will cause severe damage to the epithelium and are the main direct causes of the disease symptoms. The CDT binary toxin, present in some *C. difficile* strains, as those of RT027, is also represented. Other virulence factors (e.g. the flagellum, the S-layer) are not represented for simplicity. Shedding of the highly resistant spores will allow their accumulation into the environment and the infection of new hosts. Spores that remain in the host, in an unknown niche, are also the cause of disease recurrence.

cessation of antibiotic treatment, causing chronic intestinal disease [40]. During this period, the bacteria persisted as microcolonies and biofilm-like structures at the surface of the intestinal mucosa [40]. Bacteria within the biofilm are protected and therefore more resistant to antibiotics and oxygen [38]; conversely, spores may help shielding the biofilm. While an evolutionary link exists between sporulation and biofilm formation [41], a direct demonstration of the role of biofilms in disease and whether *C. difficile* persist in the intestine after antibiotic treatment as a biofilm, spores or both, is, however, lacking. Although the toxins, responsible for most of the disease symptoms, and spores (as a transmission/persistence factor), are the major *C. difficile* virulence factors [22, 37, 42], other virulence factors include the S-layer, fibronectin-binding proteins, flagella, fimbriae and the heat shock protein GroEL (for recent reviews, see [10, 11, 43]).

1.3. Spore formation

Spores are arguably the most resilient cellular form known to us; they are hard to eradicate and can accumulate and persist in the environment for long periods of time, without losing viability [16, 44–46]. The spore thus has a central role in the persistence of the organism in the environment, infection, recurrence and transmission of the disease [37].

Two classes of Firmicutes are able to produce endospores: the *Bacilli*, which includes the extensively studied model organism *Bacillus subtilis*, and the *Clostridia*, to which *C. difficile* belongs. The development of new tools that allow the genetic manipulation of *C. difficile*, together with transcriptomic and proteomic analysis, provided insight onto the composition and structure of the spore, and onto the spore differentiation process [47–54]. Sporulation proceeds through a series of well-defined morphological stages that culminates about 8 h after the onset of the process in the production of (usually) one dormant spore per cell (**Figure 3A**) [55–58]. The morphological steps of the sporulation process observed in *B. subtilis* are conserved in *C. difficile* [53]. Sporulation begins with a vegetative cell that contains two copies of the chromosome which becomes condensed to form a single filament stretching along the long axis of the cell. An asymmetric division then partitions the cell into a small forespore and a larger mother cell (**Figure 3A**). At this stage, the two cells lie side by side. Asymmetric division traps about 30% of one chromosome in the forespore, while the remaining of the chromosome is pumped into the forespore following division. Later, the mother cell engulfs the forespore to produce a free protoplast isolated from the external medium (**Figure 3A**). The engulfed forespore is separated from the mother cell cytoplasm by a system of two membranes that derive from the septal membranes. Next, the engulfed forespore is surrounded by two peptidoglycan (PG) layers, the primordial germ cell wall (PGCW) and the cortex, and by proteinaceous surface layers (see subsequent text). At this point, the spore becomes phase bright (**Figure 3B**) and develops full resistance to physical and chemical agents. At the end of the differentiation process, the mother cell lyses to release the mature spore. At the transcriptional level, the process is controlled by a cascade of cell type-specific alternative RNA polymerase sigma factors [55, 56, 59] (**Figure 3A**). Cell type-specific gene expression and single-cell analysis of transcription and protein localization have been monitored in *C. difficile* using oxygen-insensitive fluorescence-based reporters (reviewed in Ref. [60]).

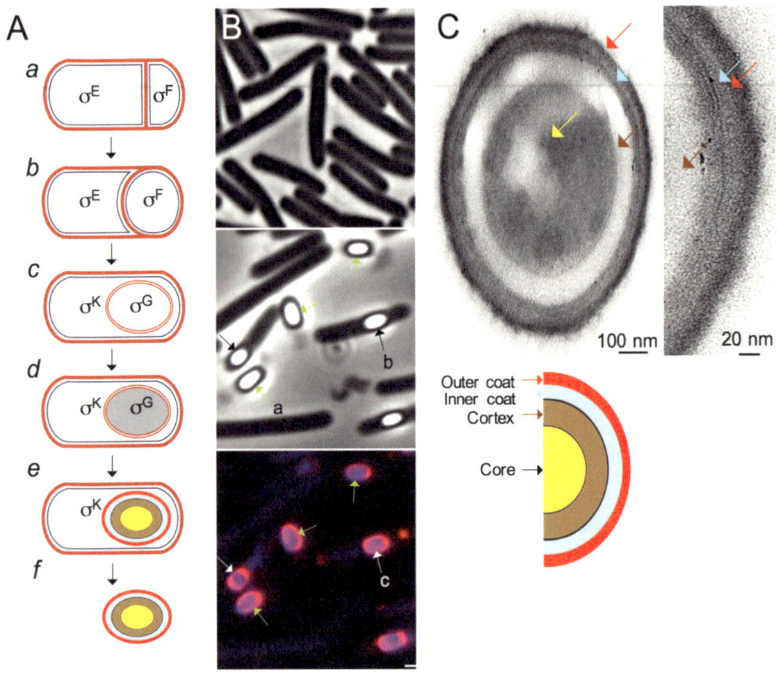

Figure 3. Sporulation in *C. difficile*. A: At the onset of the process, the rod-shaped cells divide asymmetrically to produce a larger mother cell and a smaller forespore (the future spore) (*a*). Asymmetric division involves PG synthesis within the septum. The mother cell then starts to engulf the forespore (*b*), eventually releasing it as a free protoplast inside its cytoplasm (*c*). PG polymerization contributes to the engulfment process. Following engulfment completion, the forespore is no longer in contact with the external medium and is separated from the mother cell by a system of two membranes that derive from the asymmetric division septum. Following engulfment completion, the forespore becomes visible as a phase dark body inside the mother cell (*d*). Synthesis of the primordial germ cell wall takes place from the forespore, whereas synthesis of the spore cortex PG layer is a function of the mother cell (see also insert in Figure 4). Development of full spore refractility coincides with the formation of cortex. Finally, the coat and exosporium are assembled (*e*). The spore is released into the environment through autolysis of the mother cell (*f*). B: Phase contrast microscopy and fluorescence microscopy of a *C. difficile* vegetative (top) and a sporulating culture (middle and bottom panels). For the sporulating culture, at a late stage in spore morphogenesis, the population consists of free spores (a) and refractile spores still inside the mother cell (*b*). The refractile spores exhibit most of the resistance properties of the released spores. In the bottom panel, the cells were stained with DAPI (a DNA dye) and the membrane dye FM4-64, which stains the developing spore and also the free spores (*c*). Scale bar, 1 μm. C: transmission electron microscopy (TEM) image of a thin cross section of a *C. difficile* spore. The main spore structures are labelled in the diagram. Note that an exosporium-like layer is not visible in the microscopy image, but its position, at the edge of the outer coat, is indicated in the diagram. The panel on the right shows a magnification of the spore surface. The diagram identifies the main structures or compartments normally seen by TEM. Scale bar, 0.2 μm.

1.4. Spore functional morphology

The basic endospore architecture is conserved across species. Transmission electron microscopy (TEM) shows three main concentric compartments (the core, cortex and surface layers) (**Figure 3C**) [53]. The core contains the bacterial chromosome compacted by the action of small acid-soluble spore proteins (SASPs) that can bind to the DNA altering its conformation [61]. These proteins provide resistance to damaging factors such as dry heat, UV mutagenesis, nucleases, chemicals and desiccation [62]. Immediately juxtaposed to the inner spore membrane that delimits the core is the PGCW, whose composition is similar to the vegetative cell wall and serves as a primer for the assembly of the cell wall by the newly formed cells that result from spore germination [63]. The cortex is a more external layer formed by a specialized PG, essential for the maintenance of the dehydrated state of the spore core, spore mineralization, heat resistance and dormancy [46]. While the formation of the PGCW is controlled from the forespore, the assembly of the cortex is mainly a function of the mother cell [64, 65]. Surrounding the cortex is a proteinaceous coat layer. The coat consists of an inner layer and an electron-dense outer layer (**Figure 3C**). Enzymes constitute an important part of the identified coat-associated proteins and are responsible for the enzymatic activities present at the surface of spores and which contribute to protection against chemical and physical agents (such as organic solvents, oxidative agents and UV light) [46, 50, 64]. The coat also protects the cortex layer from the action of PG-breaking enzymes, and has an important role in the interaction of spores with germinants, abiotic and biotic surfaces [64, 65]. In some pathogens, an additional layer, termed exosporium, surrounds the coat [64, 66]. The exosporium contributes to spore protection, acts as a selective permeability barrier and modulates germination through the action of associated enzymes and interactions with host cells [64, 66]. In *C. difficile,* the presence of an exosporium-like structure remains a matter of controversy; several reports suggest that this layer is fragile and easily lost [50, 67], whereas other reports indicate that it is a stable layer which is only removed by proteases and/or sonication [68–71]. The morphology of the exosporium-like layer is strain dependent; some strains have an electron-dense, compact exosporium-like layer attached to the coat, whereas others have a hair-like exosporium-like layer [58, 71, 72]. *C. difficile* has three paralogs of a collagen-like glycoprotein, BclA, which in the pathogens *B. anthracis* and *B. cereus* forms the external hair-like nap of the exosporium [66]. In these organisms, BclA forms highly stable trimeric filaments that may contribute to spore rigidity; it mediates interactions with host cells and spore uptake, and an immune-modulatory mechanism that promotes spore persistence [66]. Spores of a *C. difficile bclA1* mutant germinate faster presumably because of increased accessibility of germinants to their receptors, but fail to colonize mice; thus, proper assembly of the spore surface is essential for colonization of the mouse gastrointestinal tract [73]. While sporulation by *C. difficile in vitro* is highly asynchronous and is completed after days of incubation, sporulation genes are quickly induced following inoculation of germ-free mice [74]. Further emphasizing the importance of the spore surface layers, the most highly expressed gene *in vivo* following inoculation of germ-free mice, *cdeM,* codes for a component of the spore surface layers, and an insertional *cdeM* mutant shows impaired colonization [70, 74].

1.5. Spore germination and outgrowth

To cause disease, the dormant *C. difficile* spores must germinate in the host gastrointestinal tract. With appropriate environmental stimuli, spores initiate germination leading to the resumption of vegetative growth if sufficient nutrients are present. *C. difficile* responds to unique germinants, such as bile salts [35, 75]. While the bile salt cholate (CA) induces spore germination, another primary bile salt, chenodeoxycholate (CDCA), has been identified as a potent inhibitor of the process [35, 76]. Bile salt levels are influenced by the commensal gut microbiota. *C. scindens,* for instance, has a bile acid 7α-dehydroxylating activity, which converts CA into CDCA and has been implicated in resistance to infection by *C. difficile* [77] (see also the subsequent text) (**Figure 2**). Upon antibiotic administration, the disruption of the microbiota alters the metabolism of these two compounds and the CA concentration becomes higher than CDCA, triggering spore germination [78]. Germination induced by the CA derivative taurocholate (TA) is also enhanced in the presence of amino acids, such as glycine and histidine that act as co-germinants [35, 79]. In *C. difficile*, once germinants reach their receptors, located in the cortex region, the lytic enzymes that hydrolyse the cortex are activated [80–82]. Cortex hydrolysis allows the germ cell wall and the inner spore membrane to expand, and a massive release of calcium-dipicolinic acid from the core is observed [80, 83]. This leads to complete rehydration, enzyme activation, initiation of metabolism, macromolecular synthesis and spore outgrowth, a process during which the spore protoplast is converted into a vegetative cell [84, 85].

2. Antimicrobial therapy and the development of *C. difficile* infection

CDI is paradoxical as it occurs in the setting of antibiotic administration to treat other bacterial diseases. The disturbance of the indigenous gut microbiota by antibiotic administration is a key component, together with other risk factors, in the susceptibility to CDI.

Although the human gut microbiota is a complex ecosystem consisting of a vast number of bacteria, Archaea, viruses, yeast and fungi, the bacterial part of the microbiota is the most studied, mainly through metagenomic approaches, and is essentially formed, in healthy adults, by anaerobic bacteria belonging to the Bacteroidetes and Firmicutes phyla [86]. This ecosystem, which has coevolved with its human host, is essential to health, and is involved in many physiological functions, including metabolic transformations and regulation of immune responses [87, 88]. On the other hand, the disruption of the gut microbiota (or dysbiosis), as through antibiotic exposure, is associated with the pathogenesis of both intestinal and extra-intestinal disorders [89–91].

2.1. Effect of antibiotics on gut microbiota

Several studies in humans or in mouse models have shown that antibiotics radically alter the composition of the colonic microbiota, significantly decreasing the richness and diversity of the bacterial community, as well as its metabolic state. Dethlefsen and Relman demonstrated, by pyrosequencing of the *16S rRNA* gene, that the effect on the human gut microbiota of a 5-day treatment with ciprofloxacin was profound and rapid, with a loss of diversity and a shift in community composition occurring within 3–4 days of drug initiation [92]. In all three individuals involved in the study, the taxonomic composition of the community closely resembled its pre-treatment state by 4 weeks after the end of treatment, but several taxa failed to recover within 6 months [92]. In parallel, mouse models have shown similar long-term disturbance to the gut microbiota after the intake of antibiotics. Using pyrosequencing targeting the V6 hypervariable region of the *16S rRNA* gene, Antonopoulos and co-authors showed that an antibiotic cocktail of amoxicillin, metronidazole and bismuth induced a shift in gut microbial community structure, with an increase in Proteobacteria and a decrease in the dominant baseline microbial communities of Bacteroidetes and Firmicutes [93]. Another study performed on mice and using the same technology targeting the V1-V3 of the *16S rRNA* gene showed that a single dose of the broad-spectrum antibiotic clindamycin markedly reduced the diversity of the intestinal microbiota for at least 28 days, with an enduring loss of ~90% of normal microbial taxa from the caecum [94]. The extensive duration of the impact of clindamycin is consistent with human studies demonstrating that *Bacteroidetes* species in the faecal microbiota are reduced within 2 years following clindamycin therapy [95]. Finally, an integrated multi-omics approach, addressing the total microbiota, active microbiota, metagenome, metatranscriptome, metametabolome and metaproteome, to evaluate the changes in the faecal microbiota of a single patient after β-lactam-treatment, showed that antibiotics significantly alter the gut microbial ecology and interactions with host metabolism [96].

2.2. Antibiotics, the gut metabolic state and susceptibility to CDI

CDI is one of the gastrointestinal diseases that occur in the setting of antibiotic administration. Indeed, antimicrobial therapy is one of the main risk factors for CDI, as alteration of the gut microbiota increases the susceptibility to CDI [94, 97].

This increased susceptibility is directly correlated with the metabolic state of the altered gut microbiota. High-throughput metabolomics studies, using proteomics and multiple mass spectrometry, performed on microbiome after antibiotic treatments have shown that antibiotics reduce the levels of most products of bacterial metabolism (such as secondary bile acids, glucose, free fatty acids and dipeptides), while promoting accumulation of their precursors (oligosaccharides, sugar alcohols and primary bile acids), reflecting the modified metabolic activity of the altered gut microbiome [98, 99].

Cumulative evidence indicates that antibiotic-mediated alteration of the gut microbiome, besides reducing competitive indigenous flora, converts the global metabolic profile to one that favours *C. difficile*, both spore germination and growth of the bacteria. Indeed, *in vitro* and *ex vivo* analyses demonstrate that *C. difficile* can exploit specific metabolites that become more abundant in the mouse gut after antibiotic treatment, including the primary bile acid CA for germination, and carbon sources such as mannitol, fructose, sorbitol, raffinose and stachyose for growth [98]. Infection by *C. difficile* results in the induction of genes involved in fermentation and carbohydrate transport and metabolism [74, 100].

It is interesting to note that an abundant metabolic product resulting from bacterial growth in the gut, butyrate, induces the differentiation of colonic regulatory T cells [101]. In addition, it is well documented that this compound can decrease intestinal permeability and enhance colonic defence barriers by increasing mucin production and antimicrobial peptide levels, thus preventing host from infection [102]. Therefore, the decrease or elimination of butyrate due to microbiota dysbiosis will impair the intestinal defence barrier and increase osmotic load in the intestinal lumen, contributing to CDI susceptibility or occurrence. On the other hand, the control of *C. difficile* toxin A and B genes expression seems to be dependent on the bacterium's nutritional environment. As mentioned above, the main positive regulator of toxins expression is RNA polymerase sigma factor TcdR, whose production is influenced by various environmental signals. Regulatory proteins, such as CcpA, CodY, PrdR, SigL and Rex, and the Agr quorum-sensing system also play a role in controlling toxin gene expression in *C. difficile*, linking the metabolic and redox state of the cell to virulence [103].

The *agr1* locus is present in all sequenced strains of *C. difficile* and consists of two genes, *agrB1* and *agrD1*, that direct the production of a thiolactone, also known as the T1 signal, which accumulates extracellularly in a cell density-dependent manner [104, 105]. The T1 signal is required for *tcdA* and *tcdB* transcription, and when added to cultures, it is sufficient to cause premature expression of the toxin-encoding genes, suggesting a key role in regulating toxin production during growth [105]. At least some RT027 strains, such as R20191, have a second *agr* locus, termed *agr2*; in contrast to the *agr1* locus which carries only the genes for generation of the quorum signal, the *agr2* locus carries both the signal generation module and the genes required for signal detection (*agrC2*, coding for a sensor kinase) and transduction (*agrA2*, coding for a response regulator) [105, 106]. The *agr1* locus, however, seems essential for the generation of the T1 signal even in RT027 strains [105]. It is not known which two component systems are involved in T1 signal detection and transduction in strains lacking *agr2*. Importantly, *agrA* is required for complete expression of the toxin-encoding genes as well as the flagellar regulon, and an insertional mutant is impaired in colonization and infection in a mouse model [106]. In the *agrA2* mutant, decreased expression of the genes coding for three phosphodiesterases is likely to stimulate the degradation of c-di-GMP [106], may directly contribute to reduced expression of the gene coding for σ^D and, thus, reduced expression of *tcdA* and *tcdB*, in line with the observation that the flagellar regulon influences toxin production [107].

CcpA, in particular, binds directly to the regulatory region of several PaLoc genes, including (and with greater affinity) to *tcdR*, exerting glucose-dependent repression of toxin production; CodY, which binds branched-chain amino acids and GTP and represses the expression of many genes involved in responses to nutrient limitation, also binds directly to the *tcdR*-regulatory region [103] (**Figure 1A**).

Taking butyrate as an example, during infection, *C. difficile* appears to be able to utilize this compound [74], an activator of toxin synthesis, by a yet unknown molecular mechanism, and the genes involved in the metabolic pathway converting succinate or acetyl-CoA into butyrate are also regulated by CcpA, CodY and Rex. Thus, these metabolic regulators contribute to control the production of TcdA and TcdB by regulating the synthesis of butyrate (reviewed in Refs. [103, 108]). *C. difficile* Spo0A also contributes to the regulation of metabolism and, at least in some strains, to toxin production. Indeed, glucose uptake, glycolysis and butyrate production are downregulated in a *C. difficile spo0A* mutant [109]. That butyrate is produced by several bacteria of intestinal tract as well, suggesting that *C. difficile* turns on toxin synthesis when in the presence of other butyrate-producing species; this appears paradoxical considering that *C. difficile* is only able to colonize the colon when the normal microbiota is compromised. Discovering how the bacterium regulates butyrate metabolism and associated toxin production will likely unravel new ways of attenuating virulence, as suggested [103, 108]. In all, the direct or indirect control of the expression of the toxin-encoding genes by global metabolic regulators suggests that virulence is part of a strategy to enhance the availability of nutrient resources [103]. The link between the expression of *tcdR* and flagellar assembly, resulting from the σ^D-type promoter in the *tcdR*-regulatory region [27, 107] (**Figure 1A**), may also be viewed in this context. However, while σ^D drives the production of the toxins, it is unknown whether toxin export coincides with flagellar assembly and motility. Elevated levels of c-di-GMP promote, in several bacteria and also in *C. difficile* [110], sessile growth in detriment of motility; in *C. difficile*, elevated levels of c-di-GMP reduce the expression of the gene coding for σ^D and thus also the expression of *tcdR* and of the PaLoc genes [111] (**Figure 1A**). Finding an appropriate niche may suppress motility, and intracellular TcdA and TcdB may be only exported at this stage; the overlap between toxin production and motility, however, requires further investigation.

The link between toxin production and spore differentiation is also unclear. It is unknown whether the population of cells that produces TcdA and TcdB coincides with the population that enters sporulation, or whether toxin producers and sporulating cells represent distinct populations. It is also unknown if and to what extent motility, spore formation and toxin production overlap. In any event, some degree of coordination exists between these processes, as emphasized by the recent discovery of regulatory protein RstA [112]. RstA represses transcription of the *sigD* gene, and thus it also curtails toxin production while positively controlling sporulation initiation, regardless of the presence of a functional *sigD* gene [112].

3. The *Clostridium difficile* peptidoglycan biosynthesis pathway

Peptidoglycan, one of the components present in the bacterial cell wall, is the target of some of the more effective antibiotics known. PG is required for cell division, cell elongation and also for spore differentiation (asymmetric division at the onset of sporulation, engulfment and synthesis of the PGCW and cortex) (**Figure 3**). We provide an overview of the PG-biosynthetic pathway in *C. difficile*, as deduced from genome information, and we integrate information on the structure and synthesis of the PG macromolecule with resistance mechanisms to antibiotics that target this pathway.

PG is a heterogeneous polymer of glycan chains cross-linked by short peptides and is the only common polymer of both Gram-negative and Gram-positive cell walls. While Gram-negative cell walls are composed by a thin layer of PG located between the cell membrane and the outer membrane which is composed mainly of lipopolysaccharides, the Gram-positive cell wall, in general, has a thick layer of PG decorated by accessory polymers, such as teichoic acids [113, 114]. Secondary cell wall polysaccharides are also present in *C. difficile* but will not be covered here [115]. PG is not only essential for the preservation of cell integrity, as it confers mechanical resistance against pressure, but also has an important role in the maintenance of cell shape and anchoring of proteins and other polymers on the cell surface [116, 117].

The polysaccharide chains that form the PG are composed of alternating N-acetylglucosamine (GlcNAc) and N-acetylmuramic acid (MurNac) residues linked by β 1$\rightarrow$ 4 bonds. The MurNac residue has a stem peptide linked to the carboxyl group. The chains are cross-linked through the formation of peptide bonds between these stem peptides of alternating strands [114, 117, 118]. These stem peptides have the sequence L-Ala, D-Glu, meso-diaminopimelate (mDAP), D-Ala and D-Ala in all Gram-negative, most cyanobacteria and *Bacillus* and *Clostridia* species. In several Gram-positive species, however, mDAP is substituted by L-Lys (e.g. in

Staphylococcus aureus) [116, 119]. The biosynthesis of PG is divided into three stages: (1) synthesis of the nucleotide precursors UDP-GlcNAc and UDP-MurNAc (cytoplasmic stage), (2) synthesis of lipid-linked intermediates (cytoplasmic/membrane stage) and (3) polymerization of the PG monomers (cell surface stage) [120, 121] (**Figure 4**).

Synthesis in the cytoplasm involves the action of six Mur ligases (MurA to MurF) that catalyse the formation of UDP-MurNac-pentapeptide from UDP-GlcNAc (**Figure 4**). In the process, UDP-GlcNac is converted to UDP-MurNAc by two sequential reactions catalysed by MurA and MurB. Next, the amino acids of the stem peptide are added sequentially to the UDP-MurNAc residue through the action of MurC, D, E and F. MurC is responsible for the addition of the first amino acid which corresponds to an L-Ala. MurD recognizes the UDP-MurNac-L-Ala and adds the second amino acid (D-Glu). MurE adds the third amino acid, either mDAP or L-Lys. Finally, MurF adds the fourth and fifth as a dipeptide D-Ala-D-Ala (D-Ala-D-Ser or D-Ala-D-Lac in some vancomycin-resistant organisms; see also subsequent text), leading to the formation of UDP-MurNAc-pentapeptide [116, 117, 119, 120]. Both the Alr racemase, involved in the formation of D-Ala from L-Ala, and the Ddl ligase, involved in the formation of D-Ala-D-Ala, are inhibited by D-cycloserine (see **Figure 4**).

The membrane stage starts with the transfer of the phospho-MurNAc-pentapeptide moiety from the soluble UDP-MurNAc-pentapeptide to the membrane receptor undecaprenyl pyrophosphate (C_{55}-P, also known as bactoprenol), yielding undecaprenyl-pyrophosphoryl-Mur-NAc-pentapeptide (or lipid I). This transfer reaction is catalysed by the integral membrane protein MraY. In a second step, MurG catalyses the formation of a β 1→4 bond between MurNAc and a GlcNAc moiety from a UDP-GlcNac molecule (**Figure 4**). *In vitro* selection experiments have shown that *murG* is the site of mutations conferring increased resistance to vancomycin (see also subsequent text) [122]. The MurG-catalysed reaction produces the PG monomer undecaprenyl-pyrophosphoryl-Mur-NAc-(pentapeptide)-GlcNAc (lipid II). Flippases, the partially redundant MurJ and Amj proteins in *B. subtilis*, then translocate lipid II to the *trans*-side of the membrane [117, 123, 124]. In *B. subtilis*, a sporulation-specific protein, SpoVB, produced in the mother cell under σ^E control (**Figure 3A** and **Figure 4**), most likely fulfils the same role during spore formation [125–127]. The MurJ homologue of *C. difficile* is coded for by *CD630_10680* while a second flippase (*CD630_34980*) may be the homologue of SpoVB (**Figure 4**); whether the latter is specifically involved in spore formation and thus whether *C. difficile* relies on a single flippase for growth is presently unknown.

During the cell surface stage, the glycan strands are polymerized and peptide bridges are created between adjacent strands. Polymerization occurs through a transglycosylation reaction between the C_1 from the MurNAc residue of the nascent strand and the C_4 from the GlcNAc residue of the lipid II-linked precursor. Cross-linking of the glycan strands generally occurs between the D-Ala at position 4 of the stem peptide and the mDAP (or L-Lys) at position 3 of a stem peptide from an adjacent strand. This reaction is catalysed by the transpeptidase domain present in all PBPs that are able to cleave the D-Ala-D-Ala bond of the stem peptide, releasing the last D-Ala residue, which energizes the transpeptidase reaction [117, 128]. The undecaprenyl-pyrophosphate is translocated back to the inner side of the membrane and recycled, to receive a new UDP-MurNAc-pentapeptide molecule [121, 128] (**Figure 4**). Until

http://dx.doi.org/10.5772/intechopen.69053

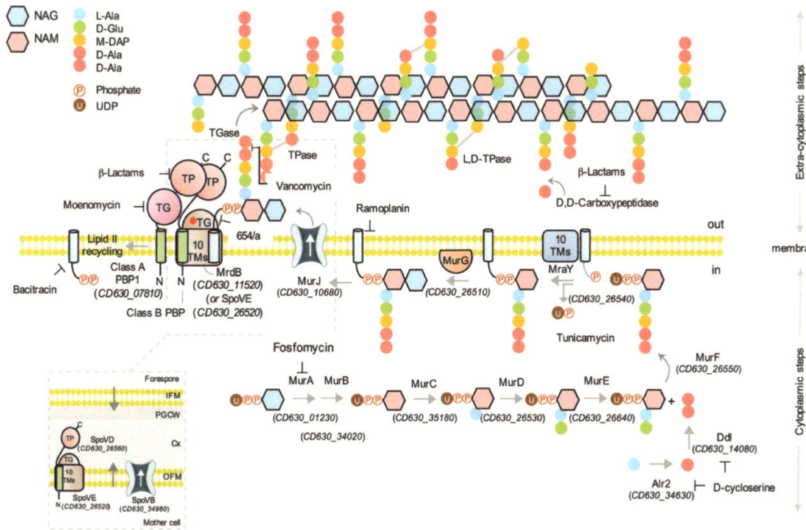

Figure 4. Overview of the *C. difficile* peptidoglycan biosynthetic pathway. The *C. difficile* counterparts of the genes known to intervene at the represented steps in the pathway are highlighted (the ORF code for strain 630Δerm is used). No cytoskeletal proteins are drawn for simplicity. The boxed complex, the core of which is formed by a SEDS-type transglycosylase (active site on the extracytoplasmic side of the membrane represented by a red dot) and a Class B PBP may function in elongation or division, depending on the functions of the represented proteins, which remain to be determined. Synthesis of the spore cortex PG follows a similar pathway but may involve sporulation-specific proteins, including a complex between a SEDS-type transglycosylase, SpoVE, and SpoVD, a Class B PBP (insert; IFM, inner forespore membrane; OFM, outer forespore membrane); SpoVB may be a mother cell-specific flippase. The direction of synthesis of the cortex (Cx) and the primordial germ cell wall (PGCW) is shown by arrows. Steps blocked by selected antibiotics are shown; the site of action of 654/a is also shown although this compound is a specific inhibitor of the SEDS protein RodA of *B. subtilis*. TM, transmembrane domain; TGase, transglycosylase domain; TPase, transpeptidase domain.

recently, the transglycosylation reaction has been almost exclusively attributed to the action of Class A PBPs (containing a transglycosylation domain; see subsequent text; see also **Figure 5A**). Some bacteria, however, lack Class A PBPs, whereas others are able to survive without their Class A enzymes [129–132]. The very recent demonstration that integral membrane proteins of the SEDS family (shape, elongation, division and sporulation) [133] have transglycosylase activity solved the paradox [134–136]. In fact, in *B. subtilis*, RodA (a SEDS protein) is a transglycosylase that associates with the complex responsible for PG synthesis during cell elongation [128, 134] (**Figure 4**). In *B. subtilis*, a RodA paralog, FtsW, has an equivalent function during cell division and a third protein, SpoVE, is specifically required for the synthesis of the spore cortex. SEDS proteins functionally cooperate with elongation- or division-specific Class B PBPs (transpeptidases). In *B. subtilis*, the direct interaction between SpoVE and SpoVD (a sporulation-specific transpeptidase) is essential for the formation of the spore cortex and spore heat resistance [125, 137] (see also **Figure 4**).

Homologues of all the enzymes involved in the cytoplasmic membrane and extracytoplasmic steps of PG biosynthesis are found in the *C. difficile* genome (**Figure 5**). Some important differences in the final structure of the PG do exist, however. Firstly, up to 93% of the GlcNAc residues are *N*-deacetylated while no modifications are found in MurNAc [115, 138, 139].

N-deacetylation of GlcNAc has been documented in other Gram-positive bacteria such as *B. subtilis, B. anthracis, Listeria monocytogenes* and *Streptococcus pneumoniae* and confers resistance to cleavage of the β 1→ 4 bonds between MurNAc and GlcNAc by a muramidase (lysozyme), a first-line defence of the host innate immune response [140, 141]. *N*-deacetylation is achieved through the action of deacetylases such as PgdA from *S. pneumoniae* [140]. In *C. difficile*, complete resistance to lysozyme involves the extracytoplasmic sigma factor σ^V, which is induced by lysozyme, and leads to further PG deacetylation [139]. Importantly, disruption of the gene coding for σ^V (*csfV*) results in a strongly attenuated mutant in a hamster model of CDI. The *pdaV* gene (*CD630_1556*) codes for a protein with homology to PG

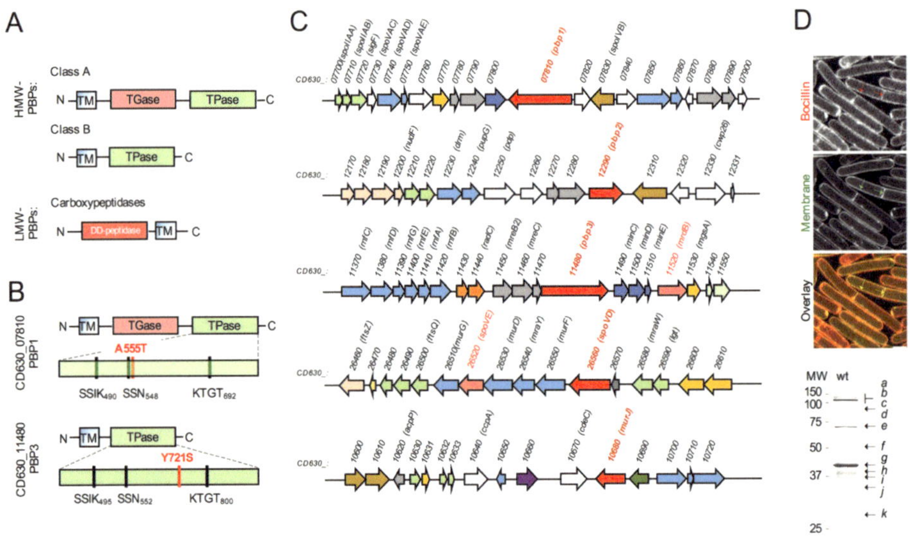

Figure 5. Class A and B penicillin-binding proteins of *C. difficile*. A: General structure of high- and low-molecular-weight (HMW and LMW) penicillin-binding proteins (PBP's). TM, Transmembrane domain; TGase, transglycosylase domain; TPase, transpeptidase domain; DD-Pep, DD-carboxypeptidase. B: The amino acid substitutions in PBP1 and PBP3 of *C. difficile* shown in red (A555T and Y721S) have been found in clones with increased resistance to imipenem, a carbapenem. The substitutions are indicated relative to the sequence found in a susceptible strain, such as the widely used laboratory strain 630Δerm. The groups of residues shown below the black lines form the TPase catalytic site, with the numbers indicating the position relative to the N-terminus of the protein. These mutations may confer resistance by decreasing the affinity of the enzyme for the antibiotic. The amino acid residues are represented in a single letter code. C: The genes coding for the four HMW Class A and B PBPs of *C. difficile* and the gene coding for the MurJ flippase and their genomic contexts. Note that the proteins most similar to the products of CD630_07810, CD630_12290 and CD630_11480 are PBP1, PBP2 and PBP3 of *B. subtilis*, respectively. CD630_26520 and CD630_26560 are labelled as *spoVE* and *spoVD*, as they occupy similar relative positions in the *dcw* cluster of *B. subtilis. B. subtilis* has two other SEDS proteins, FtsW and RodA that function specifically in cell division and cell elongation, respectively, while SpoVE is involved in the synthesis of the cortex PG during sporulation. Since *C. difficile* only has two SEDS proteins, it is likely that CD630_26520 supports cortex synthesis and either elongation or division. SpoVD is also specifically involved in the synthesis of the spore cortex in *B. subtilis* and possibly also in *C. difficile*. D: growing cells of *C. difficile* were labelled with Bocillin 650/665 Penicillin and with the membrane dye MTG and imaged by fluorescence microscopy; the bottom panel is the merge between the images collected in the red (Bocillin 650/665 Penicillin) and green (MTG/Membrane) channels. Note the labelling of both the division septa and the lateral wall of the cell. Scale bar, 1 μm. The bottom panel shows the SDS-PAGE and fluorimaging analysis of extracts prepared from the labelled cells. At least 10 bands are detected, although some may be stable proteolytic fragments of higher molecular weight species (bands *a-i*). Additional experiments are required to assign the bands detected to a specific PBP.

deacetylases and is itself under the control of σ^V; the expression of *pdaV* from an inducible promoter bypasses the requirement for *csfV* for lysozyme resistance and increased the level of *N*-deacetylated GlcNAc [139]. Thus, PdaV is a PG deacetylase, and complete resistance to lysozyme is required during infection. *C. difficile* codes for nine other putative PG deacetylases; of those, only *CD630_32570* was upregulated *in vivo* (as compared to growth *in vitro*) during infection in *C. difficile*-monoassociated mice [74]. Another characteristic of the *C. difficile* cell wall PG is that mDAP is found at the third position of the stem peptide. While most of the peptide bridges in bacteria are made between the D-Ala residue (fourth position) and the mDAP residue through the action of D-D-transpeptidases, approximately 73% of the cross-links in the *C. difficile* PG occur between mDAP residues of adjacent strands, through 3→3 bonds catalysed by L-D-transpeptidases [138, 142]. Two L-D-transpeptidase homologues, termed Ldt_{CD1} (CD630_29630) and Ldt_{CD2} (CD630_27130), were shown to participate in these reactions; a third homologue (CD630_30070) is likely to be involved because disruption of either *ldt* gene reduces but does not eliminate 3→3 cross-linking [138]. That β-lactams, but not carbapenems, inhibit D-D- but not L-D-transpeptidases may be part of the reason why *C. difficile* is tolerant to high concentrations of these antibiotics [115, 138, 142]. Nevertheless, while the L-D-transpeptidation pathway is insensitive to ampicillin, *C. difficile* is susceptible to this drug. This suggests that D-D-transpeptidation is important for the overall assembly of the PG [138], perhaps explaining the susceptibility of *C. difficile* not only to β-lactams. Thirdly, while most species have D-Ala-D-Ala at the end of the stem peptide, a significant fraction of the tetrapeptides ends with a Gly [138]. Intriguingly, while at least the prevalence of the L,D-transpeptidation pathway and the reduced presence of D-Ala-D-Ala-ending stem peptides could in principle contribute to vancomycin resistance, *C. difficile* is susceptible to this antibiotic. Moreover, *C. difficile* carries a *vanG*-type operon (*vanGCD*) most similar to that of *E. faecalis*. The ***van*** operon codes for enzymes that synthesize PG precursors ending in D-Ser or D-Lac and others that eliminate the natural (D-Ala) precursors; it thus confers vancomycin resistance [115, 142, 143]. The *van* operon of *C. difficile* is inducible by vancomycin and causes synthesis of UDP-MurNAc-pentapeptides ending in D-Ser [144]; yet, genetic ablation of the ***van*** operon in *C. difficile*, reduced the vancomycin MIC only slightly (from 1.5 to 0.75 mg/l) [143]. One possibility is that the incorporation of D-Ala-D-Ala into PG precursors is always favoured by the MurF enzyme [143]. Also noteworthy, deletion of the *dlt* operon, involved in D-alanylation of wall teichoic acids, also reduced the vancomycin MIC slightly (from 1 to 0.75 μg/ml for the strain used) [145]. Conversely, reduced susceptibility of *C. difficile* to vancomycin has been observed but the underlying mechanism is unknown [146, 147] (**Table 1**).

3.1. Penicillin-binding proteins

PBPs belong to a superfamily of acyl serine transferases that bind β-lactam antibiotics through a transpeptidase/carboxypeptidase domain which is thereby inactivated. These proteins can be divided as high-molecular-weight (HMW) PBPs and low-molecular-weight (LMW) PBPs [117, 128] (**Figure 5A**). The HMW PBPs are multimodular proteins responsible for the insertion of new molecules in the PG and cross-link formation. Generally, they contain an N-terminal cytoplasmic tail, a transmembrane anchor and two extracytoplasmic domains joined by a linker. One of the extracytoplasmic domains has transpeptidase activity responsible for the

Antibiotic	**Mechanism of action**	**Relevance in CDI**	**Resistance frequency**	**Mechanism of resistance**
Metronidazole	DNA damage after reduction of metronidazole inside the bacterial cell	Used in treatment of mild/moderate CDI	Rare; reduced susceptibility reported in frequent ribotypes	Multifactorial; 5-nitroimidazole reductase and modifications in multiple proteins involved in DNA repair, iron uptake and metronidazole reduction (putative)
Vancomycin	Inhibits cell wall synthesis by binding to the dipeptide D-Ala-D-Ala of peptidoglycan precursors	Used in treatment of severe and recurrent CDI	Rare	Mutations in *murG* (putative)
Fidaxomycin	Inhibition of RNA synthesis by binding to RNA polymerase (in a site distinct from rifamycins)	Used in treatment of severe and recurrent CDI	Rare	Mutations in *rpoB*, *rpoC* and *rarR* (reduced susceptibility in *in vitro* mutants)
MLS_B	Inhibition of protein synthesis by binding to 23S rRNA	Associated with high risk for CDI	High	Target protection by *ermB*, localized in the elements Tn5398, Tn9164 and Tn6215, or *cfr*, localized in Tn6218
Fluoroquinolones	Inhibition of DNA synthesis by binding to DNA gyrase and topoisomerase IV	Associated with high risk of CDI; resistance contributed to spread of the epidemic ribotype 027	High; associated with frequent and epidemic ribotypes	Target modification by mutations in *gyrA* and *gyrB*
Rifamycins	Inhibition of RNA synthesis by binding to RpoB	Used adjunctively for the treatment of recurrent CDI	Common; associated with frequent ribotypes	Target modification by mutations in *rpoB*
Tetracyclines	Inhibition of protein synthesis by binding to 30S ribosomal subunit	Resistance found in multiresistant isolates; resistance shared between human and swine isolates	Common	Target protection by *tetM*, carried by the elements Tn5398 or Tn916-like
Chloramphenicol	Inhibition of protein synthesis binding to the 50S ribosomal subunit	Resistance associated with prevalent ribotypes	Uncommon	Inactivation of the antibiotic by *catD*, carried by Tn4453 elements

CDI – *Clostridium difficile* infection.

Table 1. Summary of *Clostridium difficile* resistance to antibiotics and associated mechanisms.

cross-link of adjacent stem peptides. This domain is localized in the C-terminal and has three specific motifs that compose the active site: SXXK, (S/Y)XN and (K/H)(S/T)G (**Figure 5A**). The other extracytoplasmic domain is variable and allows sorting of these PBPs in two classes [128, 148]. Class A HMW PBPs are called bifunctional PBPs since they have a transglycosylase domain, which catalyses the polymerization of the new glycan strands. Transglycosylation can occur without a functional transpeptidase domain, but inactivation of the first impairs the transpeptidase activity [121, 149]. Class B HMW PBPs have instead another domain proposed to play a role in interactions with additional components of the PG synthesis machinery, such as a SEDS protein [121, 128, 149].

The LMW PBPs are DD peptidases that in the majority of the cases catalyse DD-carboxypeptidase reactions in the D-Ala-D-Ala motifs. Contrary to the HMW PBPs, these proteins are bound to the membrane through a C-terminal-located transmembrane domain or an amphipathic helix. The catalytic domain is localized in the N-terminal. The LMW PBPs play a role in regulating the degree of PG cross-linking since removal of the carboxy-terminal residue of the stem peptide prevents the cross-linking [117, 128, 144].

The number of PBPs and the proportion of the different types vary among different species and cell shapes. The rod-shaped spore-forming *B. subtilis* has 16 PBPs: four of Class A, six of Class B and six LMW PBPs. *S. aureus*, a coccus, has only four PBPs: one Class A, two Class B and one LMW. β-lactam resistance of methicillin-resistant strains of *S. aureus* (MRSA), however, stems from the acquisition of an extra, low-affinity PBP, PBP2a [117, 128, 134, 150]. It has been suggested that rod-shape species have more PBPs in order to be able to synthesize new PG not only during cell division, as in cocci, but also during cell elongation. Spore-forming species additionally have to synthesize the spore germ cell wall and cortex [150, 151].

C. difficile has nine PBPs identified of which only one is of Class A, three of Class B and five LMW PBPs (**Figure 5C** and **Table 2**). Note that the numbering of the *C. difficile* PBPs used herein is based on the nomenclature used in a recent study [152]. Strain M68, a recent representative of RT017 [153], has an additional Class B PBP, referred to as PBP5, which may have been recently acquired by horizontal gene transfer. Labelling of exponentially growing cells with a fluorescent derivative of penicillin (Bocillin 650/665 Penicillin) shows labelling of both the lateral wall of the cell and the division septum, as seen for several other rod-shaped bacteria (**Figure 5D**). Sodium dodecyl sulphate-polyacrylamide gel electrophoresis (SDS-PAGE) analysis of whole-cell extracts prepared from the labelled cells followed by fluorimaging reveals a collection of 11 labelled bands; although some of these bands may correspond to stable degradation products, the pattern is generally consistent with the genomic information (**Figure 5D**).

Like *S. aureus, C. difficile* has only one gene coding for a bifunctional PBP in a monocistronic operon (PBP1; *CD630_07810*). The genes coding for the three Class B PBPs are *CD630_12290* (PBP2), *CD630_11480* (PBP3) and *CD630_26560* (SpoVD). The gene coding for PBP2 is also monocistronic. The gene coding for PBP3 is the last gene of an operon that codes for the MreB2 and MreC components of the cell shape/elongation machinery suggesting constitutive production of the protein and a possible involvement in cell elongation. As in *B. subtilis, spoVD* is located within a region containing several other cell division and cell wall genes, the

Class	Gene	Protein	N° aa	kDa
A	*CD630_07810*	PBP1†	897	96.5
B	*CD630_12290*	PBP2†	554	62.6
B	*CD630_11480*	PBP3†	992	111.3
B	*CD630_26560*	SpoVD	659	73.2
LMW D-D-Carboxypeptidase	*CD630_05150*	-	409	45.5
LMW D-D-Carboxypeptidase	*CD630_12910*	DacF*	387	41.9
LMW D-D-Carboxypeptidase	*CD630_16270*	VanY*	268	31.7
LMW D-D-Carboxypeptidase	*CD630_21410*	-	397	44.5
LMW D-D-Carboxypeptidase	*CD630_24980*	DacF*	429	48.2

†Nomenclature according to Ref. [152].
*Nomenclature according to the NCBI (https://www.ncbi.nlm.nih.gov/gene).

Table 2. The penicillin-binding proteins of *Clostridium difficile*.

dcw cluster that also codes for the SEDS-type cortex-dedicated transglycosylase SpoVE. It is not known whether the SpoVD homologue of *C. difficile* is also cortex-specific, as in *B. subtilis*. The *dcw* gene *CD630_26520* codes for a SEDS protein and occupies the position homologous to that of *spoVE* in the *dcw* cluster of *B. subtilis* (**Figure 5C**). Since *C. difficile* only codes for two SEDS members, it is unclear whether *CD630-26500* is a functional homologue of *spoVE* and whether it accumulates a role in spore cortex synthesis with a vegetative function, in either division or elongation. Among the LMW PBPs, it is worth noting that the putative carboxy-peptidase coded for by *CD630_16270* is a homologue of proteins able to confer vancomycin resistance (see earlier text).

3.2. Bacterial shape and peptidoglycan synthesis

PG is responsible not only for resistance against physical and chemical stress but also for the maintenance of cell shape. In nature, a wide diversity of bacteria shape exists, ranging from spheres (*cocci*) to rods (*bacilli*), that has been historically used for the characterization and classification of species [154, 155]. Cell shape is defined by the different PG synthesis apparatus present in the bacteria that, through interaction with cytoskeletal elements, define the points where newly synthesized PG is inserted. Cell shape changes through the addition of antibiotics that inhibit specific complexes or by mutations that inactivate critical components of cytoskeleton or PG-biosynthetic machinery [156]. Rod-shaped bacteria have two distinct multi-protein complexes driving PG synthesis: the elongasome and divisome. The elongasome is responsible for the synthesis of PG in the lateral wall, whereas the divisome is assembled for the synthesis of the septal PG during cell division. In general, cocci only synthesize new PG during cell division [151, 154, 157].

Among elongasome components are MreB, MreC, MreD, RodA, RodZ, a Class B PBP and/or a Class A PBP. MreB is an actin homologue that assembles into short filaments that move independently perpendicular to the long axis of the cell, and is the key element that spatially governs activity of the elongasome (reviewed in Refs. [149, 158, 159]). MreC and MreD are integral membrane proteins of unknown function and RodZ connects MreB with the synthesis machinery [104, 119, 135–138].

Two actin homologues, MreB2 (*CD630_10225*) and MreB (*CD630_01270*), are found in *C. difficile*. Interestingly, the last gene of the *mreBCD* operon codes for PBP3 suggesting the participation of this Class B PBP in lateral cell wall synthesis. Elongasome complexes include a SEDS-type transglycosylase that interacts directly with a class B PBP [149, 158]. In the case of *C. difficile*, the core of the elongasome may include PBP3 and the SEDS protein might be the RodA homologue MrdB (coded for by *CD630_11520*); this gene distances 1.9 kb from the *pbp3* gene, that is, within the genetic distance range identified by Meeske et al. for SEDS/Class B PBPs pairs [134]. The only Class A PBP codified by the genome, PBP1 (*CD630_07810*), may also be part of the elongasome. Remarkably, RodZ, thought to help linking MreB to the membrane and to extracytoplasmic complexes [158], is absent in *C. difficile*.

Among the divisome components in model organisms such as *B. subtilis* and *Escherichia coli* are FtsZ, FtsA, ZipA, DivIB, DivIC (also referred to as FtsQ and FtsB in *E. coli*), FtsL, FtsW, a Class B PBP and/or a Class A PBP. FtsZ is a tubulin homologue that organizes PG synthesis during cell division [121, 149]. FtsZ is tethered to the membrane by proteins such as FtsA and ZipA. In *E. coli* and in *B. subtilis*, the formation of the Z-ring at midcell relies on two main systems. In *E. coli*, the MinCD inhibitor oscillates from pole to pole through the action of MinE, causing the overall concentration of the inhibitor to be maximal at the poles and minimal at midcell, while in *B. subtilis*, from which MinE is absent (as is the case of most Gram-positive species), MinCD is sequestered at the poles [161]. In both organisms, nucleoid occlusion only allows polymerization in areas not occupied by the chromosome (reviewed in [161]). The conjugation of the two systems restricts Z-ring formation to midcell. FtsW is a cell division-specific SEDS glycosyltransferase [151, 160–162].

Benzamides or derivatives of the alkaloid berberine are among the compounds that block FtsZ function leading to filamentation [163–165]. Importantly, benzamides showed efficacy in a mice model of systemic *S. aureus* infection [163, 164].

The main components of the divisome complex are found in the *C. difficile* genome with a genetic organization similar to *B. subtilis*. In *B. subtilis*, however, *ftsZ* is co-transcribed with *ftsA* downstream from the *mur* genes and *divIB*, whereas *divIC* is organized in an operon upstream from *spoIIE* and *ftsL* is co-transcribed with *mraW* and *pbdB* (Class B PBP). In *C. difficile*, *ftsZ* (*CD630_26460*) is downstream from *divIB* (*CD630_26500*), but surprisingly, no *ftsA* gene is found in its vicinity or elsewhere in the genome. How FtsZ is tethered to the membrane in the absence of FtsA is unknown. DivIC (*CD630_34920*) is also upstream from *spoIIE*, required for proper division and cell type-specific activation of σ^F at the onset of sporulation (see **Figure 3A**). No *ftsL* or *ftsW* genes were identified in the genome, however; as discussed above, it is possible that the *CD630_26520* gene participates both in cortex synthesis and in cell division (**Figure 4**). The absence of class B PBP and of a SEDS protein in the vicinity of these genes may indicate that PBP1 may play a crucial role also in cell division. Between the *pbp3* and *mrdB* genes is the *minCDE* operon. This genetic organization is reminiscent of *B. subtilis*, where the *mreBCD* and *minCD* genes are co-transcribed [156]. Remarkably, however, the presence of *minE* in *C. difficile* suggests that polar division is controlled through oscillation of a MinCDE complex. Finally, a gene cluster coding for three cell division proteins called MldA, B and C (midcell-localizing division proteins) can only be found in *C. difficile* and closely related species; mutants lacking MldA and MldB loose the rod-shape and daughter cells separation is inefficient [166].

In addition to the elongasome and divisome complexes, generally found in rods, spore-forming bacteria, like *B. subtilis*, usually contain a third PG-synthesizing complex that drives biogenesis of the spore cortex and the core of which is formed by SpoVE and SpoVD homologues, as discussed above [167, 168]. It is likely that such a complex also operates in *C. difficile*. Antibiotics such as bacitracin (which interferes with the dephosphorylation of C_{55}-isoprenyl pyrophosphate), fosfomycin (an inhibitor of MurA) and D-cycloserine (which inhibits the alanine racemase Alr and the D-Ala-D-Ala ligase Ddl) block PG synthesis during spore development [123, 169, 170].

Several modifications of the cortex PG, as shown by the work in *B. subtilis*, are functionally important. The cortex has a low percentage of cross-links because of the action of D-D-carboxypeptidases. Approximately 75% of the stem peptides are removed by the DacA, DacB, DacC and DacF enzymes, with DacB and DacF playing a more essential role since spores from mutant strains lacking these two proteins are unstable and show higher cross-linking, higher core water content and decreased heat resistance. *C. difficile* codes for two DacF-like proteins (CD630_12910 and CD630_24980) and mutants unable to produce these proteins may show the same type of alterations as suggested from the work in *C. perfringens* [171, 172]. Importantly, about 33% of the MurNAc residues are in the δ-lactam form, a modification that requires the concerted action of an amidase and a MurNAc deacetylase [173]. This modification allows the cortex PG to be degraded during germination, while the germ cell wall is maintained [174]. It is likely that the same modification is found in the cortex of *C. difficile* but its structure has not yet been reported.

4. Antibiotic resistance and the emergence and spreading of epidemic *Clostridium difficile* strains: historical perspectives and changing epidemiology

Virtually all antibiotics are associated with CDI, but the higher risk is linked to prolonged administration of broad-spectrum agents. Several studies using meta-analyses to examine the risk of CDI associated with the various antibiotic classes showed that the strongest and most consistent association was with clindamycin (variable odds ratio (OR): 2.86, 16.8 and 20.43), cephalosporins, particularly those of the second and third generation (variable OR: 2.23, 3.20, 4.47 and 5.68) and fluoroquinolones (variable OR: 1.66, 5.50 and 5.65) [175–177]. In one of these studies, the association between CDI risk and fluoroquinolones was modest (OR: 1.66), but the authors argued that this was not surprising since this association was more specifically related to CDI caused by the fluoroquinolone-resistant epidemic strain [177–179]. This is consistent with the fact that ciprofloxacin causes a relatively low disruption of the anaerobic gut microflora [180]. Carbapenems also increase CDI risk consistently, although with a weak association when considered alone (OR, 1.84), but stronger when included in the group of cephalosporins/monobactams/carbapenems (OR, 5.68) [175, 177]. Tetracyclines are not associated with CDI risk (variable OR: 0.91 and 0.92) [175–177].

There are therefore two effects to consider in the association between infection and antibiotics that act synergistically. One is the effect of the antibiotic on microflora imbalance, and the other is the increased risk of CDI in a patient taking an antibiotic for which the infecting strain is resistant. Indeed, once antibiotic treatment starts, infection with a *C. difficile* strain that is resistant to the antibiotic is more likely while the antibiotic is being administered due to the presence of the antibiotic in the gut. When the antibiotic treatment stops, the levels of the antibiotic in the gut diminish rapidly, but the microbiota remains disturbed for a variable period of time, depending on the antibiotic. During this time, patients can be infected with either resistant or susceptible *C. difficile*. Accordingly, mounting evidence suggests that antimicrobial resistance in *C. difficile* is a key player in the epidemiology of CDI [181]. For clindamycin, the risk of CDI associated with a clindamycin-resistant strain is increased in patients receiving this antibiotic [182]. Regarding cephalosporins, to which *C. difficile* is intrinsically resistant, the use of this antibiotic has been identified as a CDI risk factor in hospitals for the last decades [181].

More recently, there was a rise in the fluoroquinolone-associated risk concomitantly with the emergence and geographical dispersion of a fluoroquinolone-resistant *C. difficile* strain, designated BI/NAP1/027, for restriction endonuclease analysis group BI, pulse-field gel electrophoresis-type NAP1 and polymerase chain reaction ribotype (RT) 027. In addition to the high-level resistance to fluoroquinolones, RT027 strains are characterized by the increased production of both the TcdA and TcdB toxins, the presence of the binary toxin CDT and the presence of a mutation in the gene coding for the anti-TcdR anti-sigma factor TcdC [183]. Whole-genome sequencing (WGS) and phylogenetic analysis showed that this strain emerged in North America in the early 2000s, soon after acquiring a fluoroquinolone resistance mutation in *gyrA*, causing higher rates and more severe cases of CDI, and then spread widely leading to severe healthcare outbreaks also in the UK, continental Europe and Australia [183–185]. The link between the emergence of this clone and the use of fluoroquinolones is unequivocal. Indeed, fluoroquinolones were one of the most frequently prescribed antibiotic classes in North America during the late 1990s and early 2000s, so that the selective pressure for the acquisition and maintenance of fluoroquinolone resistance within healthcare settings during this period would have been at its highest [186].

Molecular-based epidemiological studies show a constant changing in the epidemiology of CDI. While in early 2000s RT027 was responsible for CDI outbreaks of increased severity, a study from 2008 analysing the epidemiology of *C. difficile* in Europe, involving a network of 106 laboratories in 34 countries, showed that RT027 accounted for only 5% of all *C. difficile* isolates, with a great diversity of ribotypes being observed [187]. Notably, in another recent multicenter study comprising 482 participating hospitals from 19 European countries, the epidemic strain RT027 was the most prevalent (19%), although distinct regional patterns of ribotype distribution were seen [188]. In parallel, other ribotypes, of reportedly increased virulence, have started to emerge.

More recently, a toxin A-negative, toxin B-positive *C. difficile* strain, from RT017, has emerged in several countries, sometimes to epidemic proportions. In a CDI surveillance study conducted in Poland between 2004 and 2006, RT017 accounted for approximately 40% of the *C. difficile* isolates studied, while in Bulgaria the occurrence of RT017 between 2008 and 2012 reached 28% [189, 190]. In the Netherlands, one hospital was affected by an outbreak caused by both RT027 and RT017 [185]. In Asia, RT017 strains are one of the most prevalent ribotypes, particularly in China, Korea and Thailand [191–193].

In Portugal, CDI surveillance based on a network of sentinel hospitals has been carried out since 2010, and showed that RT017 is one the most common ribotypes circulating in the country [194]. In particular, one RT017 clone was shown to be endemic in a hospital from 2012 until today, and a different RT017 clone has emerged in another hospital in the beginning of 2016 (our unpublished data). Despite belonging to different genetic lineages, based on multiple locus variable number tandem repeat analysis, both of these clones harbour several genetic determinants of antibiotic resistance such as *ermB*, *tetM*, and mutations in *rpoB* and *gyrA*/*gyrB*, which confer a multiresistant phenotype (see next section for details on antibiotic resistance and associated mechanisms). In addition, they were shown to be resistant to imipenem, the first antibiotic of the carbapenems class (highly resistant to β-lactamases and widely used against drug-resistant Gram-negative bacteria). Through WGS of the endemic clone, we have identified two mutations affecting the transpeptidase domain of two penicillin-binding protein genes (*pbp1* and *pbp3*; see **Figure 5A** and **B**). The mutations are therefore likely to be associated with imipenem resistance possibly by reducing the affinity of the drug to one or both proteins [152]. The emergence of resistance to carbapenems in multiresistant clones of *C. difficile* might lead to the fast spread of these strains in hospital settings, in an analogy with the initial spreading of the fluoroquinolone-resistant strains, and thus deserves urgent and continuous surveillance.

5. Overview of resistance to different classes of antibiotics and associated mechanisms in *Clostridium difficile*

As previously mentioned, antibiotics play a major role in the development of CDI. Through the disruption of the protective gut microbiota, antibiotics promote the conditions for, not only, the germination of the *C. difficile* spores (see earlier text) once the levels of antibiotic in the gut start to decrease, but also for the growth of antibiotic resistant *C. difficile* during the treatment, when there are still high levels of antibiotic in the gut. As such, *C. difficile* strains that are resistant to several classes of antibiotics will have a selective advantage for the development of CDI [20]. Both the rates of resistance in *C. difficile* and the distribution of ribotypes vary extensively between countries. However, as a result of antibiotic-selective pressure, the most common ribotypes, which include the epidemic strains, are usually the ones presenting the highest rates of resistance [146, 188, 195, 196]. Antibiotic resistance in bacteria can be due to mutations in specific genes or due to the presence of genes acquired by horizontal gene transfer. The mechanisms of resistance are then divided into three main categories: (i) modifications of the antibiotic target, (ii) inactivation of the antibiotic and (iii) reduction of the intracellular concentration of the antibiotic [197].

5.1. Metronidazole, vancomycin and fidaxomicin

The surveillance of *C. difficile* susceptibility to the antibiotics used for CDI treatment, that is, metronidazole, vancomycin and fidaxomicin [198], is crucial. Resistance to metronidazole and vancomycin has been seldom observed but reduced susceptibility is now being reported more often [195]. In two recent studies, the MIC_{90} (the minimum concentration of metronidazole necessary to inhibit 90% of the isolates) of metronidazole was of 2 mg/L for both European and US strains, while for vancomycin it was of 2 and 4 mg/L, respectively [146, 199]. Reduced susceptibility to metronidazole has been observed in some of the most frequent RTs, including RT027, RT001 and RT106 [146, 200]. Heterogeneous and unstable resistance to metronidazole has also been described, in which cases the resistance was only observed in primary fresh isolates or after exposing the thawed isolates to low concentrations of the antibiotic, also depending on the methodology used; this indicates that these heteroresistant populations might go undetected during routine susceptibility testing [201]. Reduced susceptibility to vancomycin was also observed in two emergent RTs (RT018 and RT356) in the pan-European survey [146] and in the epidemic RT027, among US isolates [147]. Although the clinical impact of reduced susceptibility to metronidazole in CDI is still not clear, there is evidence of a correlation between reduced susceptibility and recurrent CDI, which is supported by the low concentration of metronidazole observed in the gut, unlike vancomycin which is detected at high concentrations [202, 203]. As discussed above, the mechanisms of resistance to metronidazole and vancomycin in *C. difficile* remain to be clarified (but see also subsequent text).

Metronidazole enters the bacterial cell by passive diffusion as an inactive prodrug. It is then reduced into its cytotoxic active form through the transfer of an electron to the nitro group of the drug, forming a nitroso-free radical, which interacts with DNA, inflicting DNA damage and inhibiting synthesis, ultimately leading to cell death [204, 205]. Some studies point to a multifactorial mechanism of metronidazole resistance in *C. difficile*, which includes (i) the activity of a putative 5-nitroimidazole reductase, which converts the nitro group of the prodrug into its non-toxic amine derivative; (ii) the elevated expression of proteins involved in DNA repair, such as the UvrABC excinuclease, exodeoxyribonuclease, endonuclease III, endonuclease IV and DNA mismatch repair protein; (iii) alterations in proteins related to iron uptake, such as an increased expression and mutations in the ferric uptake transcriptional regulator Fur [206, 207], which plays a central role in iron homeostasis by controlling the expression of a regulon that includes genes involved in the mitigation of oxidative stress and in redox metabolism [207]; in *Helicobacter pylori,* point mutations in *fur* have been shown to increase metronidazole resistance by reducing the binding affinity of the modified Fur protein to the promoter region of the superoxide dismutase encoding gene *sodB* and hence rendering Fur unable to efficiently repress transcription of *sodB*; high levels of SodB, in turn, allow *H. pylori* to counteract the oxidative stress generated by the activated metronidazole [208–210]; and (iv) alterations in the metabolic pathway involving pyruvate-ferredoxin oxidoreductase (Pfo), responsible for metronidazole reduction; a reduced concentration of Pfo, for example, will likely lead to a less efficient activation of metronidazole inside the bacterial cell [211].

Vancomycin inhibits bacterial cell wall synthesis by binding to the D-Ala-D-Ala dipeptide of the peptidoglycan precursor before cross-linking of adjacent peptidoglycan strands. The *in vitro* selection of isolates with reduced susceptibility to vancomycin showed the acquisition of a mutation causing a P108L substitution in *murG*; since MurG converts lipid I to lipid II, a target of vancomycin, it is conceivable that alterations in this pathway might affect the activity of vancomycin (see earlier text; **Figure 4**). Other mutations found included a stop codon in an RNA/single-stranded DNA exonuclease (*CD630_36590*), a single amino acid deletion in the *sdaB*-encoded L-serine dehydrogenase, and a missense mutation causing a D244Y substitution in *rpoC*, coding for the β′subunit of RNA polymerase. How mutations in *murG, sdaB* and *rpoC* and *dlt* operon contribute to the reduced susceptibility to vancomycin is still unknown [122] (**Table 1**).

Fidaxomicin blocks an initial step in transcription by RNA polymerase. Binding of the RNA polymerase holoenzyme to a promoter results in the formation of a closed complex, which is isomerized to an open promoter complex through opening of the double-stranded DNA; the melted region, or transcription bubble, extends approximately from positions −12 to +2 relative to the transcription start site [212]. Fidaxomicin inhibits transcription initiation if added before the stable holoenzyme/promoter open complex is formed, in contrast with antibiotics that inhibit RNA elongation such as the rifamycins [212]. *In vitro* selection for resistance led to the identification of a mutation in the *rpoB* gene (coding for the β subunit of RNA polymerase) causing the single amino acid substitutions Q1074K or Q1073R [122, 213]. Substitutions in the β′subunit have also been linked to resistance (R337A) or reduced susceptibility to fidaxomicin (two simultaneous substitutions, Q781R and D1127E) [213, 214]. Another mutation associated with reduced susceptibility to fidaxomicin, selected *in vitro,* is a frameshift mutation in the *CD630_22120* gene coding for a homologue of the multidrug resistance-associated transcriptional regulator MarR; the role of this regulatory protein in fidaxomicin resistance is still unclear [122] (**Table 2**). Resistance to fidaxomicin has only been observed in one isolate with a MIC of 16 mg/L [215], being that usually all isolates are inhibited at concentrations of ≤1 mg/L [146, 199, 216].

5.2. Clindamycin

Clindamycin is one of the antibiotics associated with an increased risk of CDI [177], and resistance to this antibiotic is one of the most common in *C. difficile*. Clindamycin resistance is often found in >30% of the isolates worldwide and frequent RTs tend to exhibit higher rates of resistance. Furthermore, clindamycin resistance is frequently found in multidrug-resistant isolates in association with resistance to fluoroquinolones and rifampicin [146, 192, 199, 216–219]. Resistance to clindamycin, a lincosamide that inhibits bacterial protein synthesis by binding to the bacterial 23S rRNA, is generally due to the presence of *ermB* (erythromycin resistance methylase gene). The rRNA adenine N-6-methyltransferase encoded by this gene methylates the adenine at position 2058 of 23S rRN, which prevents binding of the MLS_B (macrolide, lincosamide and streptogramin B) antibiotics. This gene is found in mobile elements, such as the transposons Tn5398, which contains two copies of *ermB*, Tn6194 and Tn6215. These elements are horizontally transferred between *C. difficile* strains and also to and from other genera [220]. Even so, there is a significant proportion of clindamycin-resistant *C. difficile* strains, which are *ermB*-negative [218]. Recently, a *cfr*-like gene was found to confer resistance to multiple antibiotics in *C. difficile*, including clindamycin/erythromycin, linezolid and chloramphenicol/florfenicol. Like *ermB*, this gene is found in a transposon, Tn6218, and confers resistance through modification of the bacterial 23S rRNA at position A2503 [221, 222].

5.3. Fluoroquinolones

Fluoroquinolones play a major role in the paradigm of CDI. Resistance to this class of antibiotics, which inhibit bacterial DNA synthesis by binding to the type II topoisomerases DNA gyrase and topoisomerase IV [223], was associated with the worldwide spread of the epidemic RT027 in the 2000s and rapidly became a prominent risk factor for CDI, as previously described [224]. Since then, fluoroquinolones resistance has been reported worldwide with increasing rates and is frequently found in common and epidemic strains, such as those of RT027 (in which the resistance is almost ubiquitous), RT017 and RT018 [146, 192, 225]. The rates of resistance vary considerably between countries; however, this is likely due to differences in ribotype diversity, since the countries with higher diversity of ribotypes tend to exhibit lower rates of fluoroquinolones resistance and vice versa [146]. This same trend was observed in Portugal when comparing ribotype diversity and resistance rates between regions [194]. Fluoroquinolone resistance in *C. difficile*, as a result of antibiotics use selective pressure, is associated with well-known mutations in the quinolone-resistance determining region of DNA gyrase subunits *gyrA* and *gyrB*. Mutations causing a T82I substitution in GyrA are the most commonly identified in *C. difficile*, and are found in ribotypes such as RT027 and RT017 [194, 218, 226]. Importantly, resistance to fluoroquinolones is likely to be maintained even without antibiotic pressure, as it does not impose a fitness cost in *C. difficile* [227].

5.4. Rifamycins

Rifamycins are a class of antibiotics that inhibit bacterial RNA synthesis by binding to the β subunit of RNA polymerase (coded for by the *rpoB* gene) [228]. Two members of this class, rifampicin and rifaximin, have been used adjunctively for the treatment of recurrent CDI, despite not being recommended due to the absence of evidence supporting their efficacy (vs. monotherapy) [198, 229, 230]. In fact, high rates of rifampicin resistance have been identified in epidemic strains [231]. In a recent study, rifampicin resistance (13.4%) was reported in 17/22 European countries, and although it was observed in several RTs, it was mostly associated with frequent RTs, such as RT027, RT018 and RT356 [146]. Also, in Portugal, all the RT017 strains, a predominant type [194], show high resistance to rifampicin (unpublished data). It has been demonstrated that *in vitro* susceptibility to rifampicin is predictive of rifaximin susceptibility [228] and the emergence of rifamycin resistance during a recurrent *C. difficile* infection following rifaximin treatment has been clearly demonstrated [232]. As in other bacteria, resistance to this class of antibiotics in *C. difficile* results from specific substitutions in, or near to, the residues in the β subunit of RNA polymerase that interact with rifamycins. The substitution R505K is the most frequently found in *C. difficile*, either alone or along with other substitutions, such as the frequently found H502N. Furthermore, the resistant isolates do not seem to have a clonal origin but rather appear to arise independently as the result of antibiotic pressure [218, 228].

5.5. Tetracyclines

Tetracyclines inhibit protein synthesis in bacteria by binding to the 30S subunit of the bacterial ribosome, thereby preventing the association of aminoacyl-tRNA [233]. Tetracycline resistance is a relatively common trait in *C. difficile*. The resistance rates vary widely between countries, with some presenting high rates but the majority having <10%-resistant isolates [234, 235]. The exposure to tetracyclines, however, does not seem to be associated with an increased risk of CDI [177]. Even so, resistance to tetracycline is usually found in multidrug-resistant isolates, such as those of RT012 and RT017 [218, 234, 235], and is, in most strains, associated with the presence of the *tetM* gene, which codes for a ribosome protection protein. This gene is carried by transposons Tn5397, which is found in RT012, and Tn916-like, which is found in RTs such as RT017 and RT078 [236]. TetM confers protection by binding to the ribosome in the proximity of the tetracycline-binding site and hence dislodging and preventing binding of the tetracyclines to their target [237]. Although less frequent, another *tet* gene, *tetW*, which also codes for a ribosomal protection protein, has been found in tetracycline-resistant *C. difficile* isolated from humans and animals that also harboured the *tetM* gene [218, 238]. Furthermore, the co-presence of *tetM* and *ermB* in a Tn916-like element presumably originated by the recombination of different elements has also been described [239].

Tetracyclines are the most used antibiotics for veterinary purposes [240]. Notably, the high homology between *C. difficile* RT078 isolates from human and swine has been emphasized by the common tetracycline resistance determinants found in both groups, strongly suggesting a zoonotic spread of *C. difficile* and resistance [241].

5.6. Chloramphenicol

Chloramphenicol inhibits bacterial protein synthesis by binding to the A-site of the 50S-ribosomal subunit [242]. Resistance to chloramphenicol is relatively uncommon in *C. difficile*, and is found in only 3.7% of the European isolates with a marked variation between countries that seem to reflect a localized association with specific RT, namely RT001, that apparently spread after acquiring resistance to chloramphenicol [146]. Resistance to this antibiotic is usually due to the presence of *catD*, which encodes a chloramphenicol acetyltransferase that catalyses the transfer of an acetyl group from acetyl-CoA to chloramphenicol, thus rendering the antibiotic unable to bind to the ribosome [243], and is carried by a Tn4453 element [244]. Another gene, *cfr*, has been found to confer resistance to chloramphenicol and also to other classes of antibiotics in *C. difficile*, as already mentioned earlier (see section "*Clindamycin*").

5.7. Multidrug resistance in *C. difficile*

In 2005, 82 of 316 European *C. difficile* isolates were resistant to at least three antibiotics. The predominant phenotype among these strains, which mainly belonged to RT001, RT017 and RT012, was multiple resistances to clindamycin, erythromycin, moxifloxacin and rifampicin [218]. More recently, 27.5% of RT027 strains in the US were also multidrug resistant [225]. In China, 73.3% of the strains were multidrug resistant and mainly included strains from RT017, which is predominant in Asian countries, as discussed earlier [234]. Similarly, in 2012, 85.5% of Polish strains were multiresistant, being all strains from RT027, RT176 (related to RT027), RT012 and RT046 [217]. The recent pan-European study also associates specific RT with multidrug resistance, including RT027, RT001 and RT017, which are common RT in many countries, but also RT018 and RT356 that are predominant in Italy. Another highly resistant RT, RT356, also found in Italy, is resistant to rifampicin, moxifloxacin, clindamycin, imipenem and chloramphenicol [146]. Similarly, most of RT017 strains isolated in Portugal are resistant to moxifloxacin, clindamycin, erythromycin, rifampicin, tetracycline and imipenem (our unpublished data; see also earlier text and **Figure 3B**); it seems possible that the resistance to imipenem arose due to the intense use of carbapenems in Portugal in the last decade [245]. The recently described *cfr* gene in *C. difficile*, to which we alluded to earlier, is another concern since it provides multidrug resistance and the extent of its dissemination in *C. difficile* isolates is still unknown [221]. Notably, most of the ribotypes associated with multidrug resistance are epidemic and/or associated with increased CDI severity, which hints at multidrug resistance constituting a selective advantage for the spread and infective potential of these strains. Overall, these data indicate a clear association between prevalent RT and multidrug resistance, in line with the view that antibiotic pressure drives the evolution of these strains.

6. Novel strategies for treatment and prevention of *C. difficile* infection

Strategies to neutralize *C. difficile* target all different stages in the organism's life cycle. Chlorine-based disinfectants are used to reduce the load of spores in contaminated surfaces in healthcare facilities [246]. Triggering germination has been proposed as a method to enhance the killing of spores that accumulate in healthcare facilities, in combination with UV-C or disinfectants [247]. The recent demonstration of sporicidal activity of ceragenin CSA-13 is worth mentioning. Ceragenin CSA-13 is a cationic compound that mimics endogeneous antimicrobial peptides and shows a wide spectrum of bactericidal activity [248]. When incubated with *B. subtilis* spores, ceragenin CSA-13 appears to disrupt the inner spore membrane, causing the release of Ca^{2+}-dipicolinate from the core and loss of spore viability [248]. Although likely, activity against spores of *C. difficile* was not reported.

Antimicrobial stewardship is an important aspect in strategies designed to prevent and control CDI outbreaks, in addition to infection control and containment measures and environmental decontamination [229]. The importance of antibiotic stewardship practices is exemplified and stressed here by the identification of the *pbp1* and *pbp3* alleles in imipenem-resistant strains of *C. difficile* (see earlier; **Figure 5B**). Other reviews provide a detailed discussion on measures for the treatment and prevention of CDI [9, 229]. 'Breakthrough' approaches in the treatment and prevention of CDI fall into three main areas [9]: (i) antibiotic therapies, (ii) biotherapeutics and (iii) immunological therapies [9]. We review here some of the approaches in each of these broad areas, and also refer to the effect of diet in the control of *C. difficile*.

6.1. Antibiotics

As the broad-spectrum metronidazole and vancomycin have a considerable impact on the microbiota and show high recurrence rates, there has been an intense search for new antimicrobials of narrower spectrum. The narrow-spectrum antibiotic fidaxomicin, approved by the Food and Drug Administration (FDA) in 2011, is an example. It is a macrocyclic non-systemic antibiotic shown to be highly selective against *C. difficile* and superior to vancomycin in eliciting a sustained clinical response, that is, cure without disease recurrence [249]. Several aspects of fidaxomicin action may contribute to its efficacy. *In vitro*, the drug prevents toxin production [250], and when added to cultures at the onset of stationary phase, fidaxomicin also prevented entry into sporulation [251]. It also prevented the outgrowth of cells from germinated spores (it did not prevent the initiation of spore germination, however) [252]. The inhibition of spore outgrowth and of sporulation may be the main factors contributing to the efficacy of fidaxomicin in suppressing disease recurrence and transmission (**Figure 2**).

Other narrow spectrum antimicrobials are undergoing clinical trials (www.clinicaltrials.org). Surotomycin, also called CB-183,315, is a lipopeptide antibiotic structurally related to daptomycin which is currently in phase III clinical development. As is the case for daptomycin, Surotomycin may work by dissipating the membrane potential [253]. *In vitro* studies showed activity of Surotomycin against *C. difficile* isolates with elevated MICs for metronidazole, moxifloxacin and vancomycin but lack of activity against Enterobacteriaceae and species of the *Bacteroides fragilis* group [254, 255], suggesting that it may not cause significant disruption of the microbiota.

Another example is Cadazolid. This drug is primarily a protein synthesis inhibitor, but the molecule also includes a fluoroquinolone moiety that acts as a weak inhibitor of DNA synthesis at much higher concentrations of the drug [256]. The addition of Cadazolid to stationary phase cultures inhibits the production of the TcdA and TcdB toxins and spore formation, while in both the hamster and mouse models Cadazolid was as effective as vancomycin [257]. Trials suggest that Cadazolid may be as effective as vancomycin, but with lower recurrence rates [9].

Ridinilazole [2, 2'-bis(4-pyridyl)3H,3'H 5,5'-bibenzimidazole] was more active than fidaxomicin, metronidazole and vancomycin against *C. difficile*; it was less effective against other intestinal bacteria, including the *B. fragilis* group, and Gram-positive species of *Bifidobacteria*, among others [258, 259]. Phase II trials are under way to investigate the efficacy of Ridinilazole in comparison to fidaxomicin and vancomycin. The mechanism of action of Ridinilazole has not been described. The addition of Ridinilazole to *C. difficile* cultures at concentrations below the MIC results in cell filamentation, however, suggesting that it may act as a cell division inhibitor [260].

CRS3123 (formerly REP3123) is a synthetic diaryldiamine that inhibits methionyl-tRNA synthetases and protein synthesis in Gram-positive bacteria [261]. *In vitro*, CRS3123 inhibits the growth of *C. difficile* and blocks toxin and toxin production when added to stationary phase cultures; CRS3123 demonstrated good potency against *C. difficile*, but was much less active against other bacteria of the normal intestinal flora [261–263]. It also proved superior to vancomycin in the hamster model of CDI [263]. Clinical trials to determine the efficacy and pharmacokinetics of CRS3123 have been completed but no results are yet posted.

The newly discovered SEDS-type transglycosylases, which appear as promising targets for new antibiotics, are worth noting. Screening of a library of actinomycete strains against a *B. subtilis* strain lacking all four Class A PBPs (transglycosylases/transpeptidases) led to the identification of a compound, 654/A, that caused large zones of inhibition on plates of the indicator strain (but not of the congenic WT) and morphological defects characteristic of mutants impaired in cell wall synthesis [136]. Consistent with the possibility that 654/A targets a SEDS transglycosylase, not only it acted synergistically with moenomycin, an inhibitor of the transglycosylase domain of Class A PBPs, but overproduction of the elongation-specific SEDS protein RodA mitigated its effects [136]. 654/A was also active against *S. aureus*. Activity against *C. difficile* was not reported.

6.2. Bacteriotherapies

Among the class of bacteriotherapies are the use of probiotics, spores of non-toxinogenic strains, faecal microbiota transplantation (FMT) and precise manipulation of the gut microbiota. While a number of probiotic formulations are under clinical testing, evidence for their efficacy in the treatment of CDI is presently controversial [9]. FMT, on the other hand, which relies on the disruption of dysbiosis in patients undergoing antibiotic treatment, has been used with a success rate of over 90% [9]. Difficulties associated with FMT, such as poor reproducibility, availability of the material and patient acceptance, could in principle be solved by the identification of the bacteria in faecal material responsible for the beneficial effect. The isolated strains could be produced and formulated under controlled conditions and used for therapeutic interventions. This quest is supported by several findings. Gut dysbiosis could be disrupted by a mixture of six intestinal bacteria (including S. *warneri*, *Enterococcus hirae*, *Lactobacillus reuteri*, and three novel species of *Anaerostipes*, *Bacteroidetes* and *Enterohabdus*), and this mix could clear infection by a RT027 strain in mice [40]. Also, a defined mixture of 17 gut-indigenous strains, enriched in Clostridia, isolated from the faeces of healthy humans, induced the differentiation of gut-regulatory T cells [264, 265]. A more defined intervention is illustrated by the finding that the administration of *C. scindens*, both in mice and in humans, enhances resistance to infection in a secondary bile acid-dependent manner [77]. While the depletion of the bile acid-hydroxylating activity of *C. scindens* caused by antibiotics, and the decreased resistance to CDI clearly illustrates the link between antibiotics, bile acid metabolism and susceptibility to CDI, restoration of secondary bile metabolism may be a key mechanism for the success of FMT in treating recurrent CDI [266, 267]. While these interventions rely on the bacteria that are transferred to the compromised host, a very recent report indicates that the transfer of sterile filtrates from donor faeces to patients with a diagnostic of CDI is sufficient to eliminate symptoms [268]. The authors of this study suggest that bacterial components, metabolites or perhaps phages mimic or substitute for many of the effects of FMT [268].

Also in the category of the biotherapeutic approaches to control and prevent CDI is the oral administration of spores of non-toxinogenic strains. A strain, NTCD-M3, isolated at high frequency from hospitalized patients asymptomatically colonized, was found to lack the toxin-encoding genes [269]. Spores produced by NTCD-M3 were given orally to patients under metronidazole or vancomycin treatment for their first episode of CDI or first CDI recurrence [270, 271]. Colonization was a function of the number of spores given daily. Recurrence was lower in patients receiving spores relative to the control group and correlated with colonization. Colonization was transient and lost after 22 weeks, perhaps because of the recovery of the gut microbiota [270, 271]. Presumably, recurrence was prevented because NTCD-M3 outcompeted toxinogenic strains and prevented further colonization by the latter strains. The exact mechanism, however, is not known. So far, the transfer of the PaLoc from toxinogenic to non-toxinogenic strains has only been detected *in vitro* [272].

6.3. Vaccines

Antibodies directed against the receptor-binding domains of the TcdA and TcdB toxins prevent binding to their receptors and confer protective immunity against CDI in animal models and protection against recurrent disease in humans [10, 11]. Three vaccine candidates are currently under development for CDI, all of which involve parenteral delivery of toxoids. Sanofi has a toxoid vaccine composed of partially purified and formalin-inactivated TcdA and TcdB toxins [273, 274]. The Pfizer vaccine contains mutant forms of the TcdA and TcdB toxins with mutations thought to abrogate glucosyltransferase and auto-protease activities; residual activity of the toxins, however, required pre-incubation of the antigens with specific antibodies or formalin [275]. The Valneva vaccine contains a recombinant fusion protein between the receptor-binding domains of both TcdA and TcdB which induced neutralizing levels of serum antibodies to both toxins and reduced disease severity while conferring significant protection against a lethal dose of *C. difficile* spores in hamsters [276].

In an alternative strategy, *B. subtilis* spores were used as a delivery vehicle for the carboxy terminal repeat domains of TcdA and TcdB fused to surface-exposed spore coat proteins [277]. Oral immunization with spores displaying the TcdA repeat domain alone conferred protection against challenge with a *C. difficile* strain producing both toxins, and vaccinated animals survived reinfection. Mucosal immunization was required to generate secretory IgA and the local production of these neutralizing polymeric antibodies correlated with protection [277]. A trial to assess safety and immunogenicity of a spore-based vaccine (CDVAX) started on 1 January 2017.

6.4. Diet

Finally, recent work has highlighted a role for the diet in the prevention of CDI. Zn has a role in modulating the diversity of the microbiota: mice fed with a high Zn diet showed decreased microbiota diversity, as opposed to mice on a low Zn diet [278]. A Zn-binding protein, Calprotectin added to *C. difficile* cultures, prevented growth in a Zn-dependent manner (as a mutant deficient in Zn binding did not prevent growth), and Calprotectin-deficient mice showed decreased survival and increased disease severity following challenge with an RT027 strain [278].

Acknowledgements

This work was supported by internal funding of the National Institute of Health, Dr Ricardo Jorge and by Fundação para a Ciência e a Tecnologia (FCT) (www.fct.pt) through grant Pest-C/EQB/LA0006/2011 to AOH and programme IF (IF/00268/2013/CP1173/CT0006) to MS. ALM was supported by a PhD fellowship (PD/BD/105738/2014) from the FCT.

Author details

Joana Isidro[1†], Aristides L. Mendes[2†], Mónica Serrano[2], Adriano O. Henriques[2*] and Mónica Oleastro[1*]

*Address all correspondence to: aoh@itqb.unl.pt and monica.oleastro@insa.min-saude.pt

1 National Institute of Health Dr Ricardo Jorge, Lisbon, Portugal

2 Instituto de Tecnologia Química e Biológica António Xavier, Oeiras, Portugal

† These authors contributed equally to this work

References

[1] Browne HP, Forster SC, Anonye BO, Kumar N, Neville BA, Stares MD, et al. Culturing of "unculturable" human microbiota reveals novel taxa and extensive sporulation. Nature. 2016;**533**(7604):543-546.

[2] Yutin N, Galperin MY. A genomic update on clostridial phylogeny: Gram-negative spore formers and other misplaced clostridia. Environmental Microbiology. 2013 Aug;**15**(10):2631-2641

[3] Lawson PA, Citron DM, Tyrrell KL, Finegold SM. Reclassification of *Clostridium difficile* as *Clostridioides difficile* (Hall and O'Toole 1935) Prévot 1938. Anaerobe. 2016;**40**:95-99

[4] Hall I, O'Toole E. Intestinal flora in new-born infants with a description of a new pathogenic anaerobe, *Bacillus* difficilis. American Journal of Diseases of Children. 1935 Feb;**49**(2):390-402

[5] Tedesco F, Barton R, Alpers D. Clindamycin-associated colitis: A prospective study. Annals of Internal Medicine. 1974 Oct;**81**(4):429-433

[6] Bartlett JG, Onderdonk AB, Cisneros RL, Kasper DL. Clindamycin-associated colitis due to a toxin-producing species of Clostridium in hamsters. Journal of Infectious Diseases. 1977 Nov;**136**(5):701-705

[7] Bartlett JG, Chang TEW, Gurwith M, Gorbach SL, Onderdonk AB. Antibiotic-associated pseudomembranous colitis due to toxin-producing clostridia. New England Journal of Medicine. 1978 Mar;**298**(10):531-534

[8] George W, Goldstein EC, Sutter V, Ludwig S, Finegold S. Ætiology of antimicrobial-agent-associated colitis. Lancet. 1978;**311**(8068):802-803

[9] Kociolek LK, Gerding DN. Breakthroughs in the treatment and prevention of *Clostridium difficile* infection. Nature Reviews Gastroenterology & Hepatology. 2016;**13**(3):150-160

[10] Abt MC, McKenney PT, Pamer EG. *Clostridium difficile* colitis: Pathogenesis and host defence. Nature Reviews Microbiology. 2016 Oct;**14**(10):609-620

[11] Smits WK, Lyras D, Lacy DB, Wilcox MH, Kuijper EJ. *Clostridium difficile* infection. Nature Reviews Disease Primers. 2016 Apr 7;**2**:16020

[12] Dubberke ER, Olsen MA. Burden of *Clostridium difficile* on the healthcare system. Clinical Infectious Diseases. 2012 Aug 1;**55**(suppl_2):S88–S92

[13] Lessa FC, Mu Y, Bamberg WM, Beldavs ZG, Dumyati GK, Dunn JR, et al. Burden of *Clostridium difficile* infection in the United States. New England Journal of Medicine. 2015 Feb 26;**372**(9):825-834

[14] Jones AM, Kuijper EJ, Wilcox MH. *Clostridium difficile*: A European perspective. Journal of Infection. 2013 Feb;**66**(2):115-128

[15] Britton RA, Young VB. Interaction between the intestinal microbiota and host in *Clostridium difficile* colonization resistance. Trends in Microbiology. 2012;**20**(7):313-319

[16] Carroll KC, Bartlett JG. Biology of *Clostridium difficile*: Implications for epidemiology and diagnosis. Annual Review in Microbiology. 2011;**65**:501-521

[17] Lyras D, Adams V, Lucet I, Rood JI. The large resolvase TnpX is the only transposon-encoded protein required for transposition of the Tn4451/3 family of integrative mobilizable elements. Molecular Microbiology. 2004;**51**(6):1787-1800

[18] Hussain HA, Roberts AP, Mullany P. Generation of an erythromycin-sensitive derivative of *Clostridium difficile* strain 630 (630Δerm) and demonstration that the conjugative transposon Tn916ΔE enters the genome of this strain at multiple sites. Journal of Medical Microbiology. 2005;**54**:137-141

[19] Sebaihia M, Wren BW, Mullany P, Fairweather NF, Minton N, Stabler R, et al. The multi-drug-resistant human pathogen *Clostridium difficile* has a highly mobile, mosaic genome. Nature Genetics. 2006;**38**(7):779-786

[20] Rupnik M, Wilcox MH, Gerding DN. *Clostridium difficile* infection: New developments in epidemiology and pathogenesis. Nature Reviews Microbiology. 2009;**7**(7):526-536

[21] Dawson LF, Valiente E, Wren BW. *Clostridium difficile* — A continually evolving and problematic pathogen. Infection, Genetics and Evolution. 2009;**9**:1410-1417

[22] Voth DE, Ballard JD. *Clostridium difficile* toxins: Mechanism of action and role in disease. Clinical Microbiology Reviews. 2005;**18**(2):247-263

[23] Govind R, Dupuy B. Secretion of *Clostridium difficile* toxins A and B requires the holin-like protein TcdE. PLoS One. 2012;**8**(6):e1002727

[24] Olling A, Seehase S, Minton NP, Tatge H, Schröter S, Kohlscheen S, et al. Microbial pathogenesis release of TcdA and TcdB from *Clostridium difficile* cdi 630 is not affected by functional inactivation of the tcdE gene. Microbial Pathogens. 2012;**52**(1):92-100

[25] Mani N, Dupuy B. Regulation of toxin synthesis in *Clostridium difficile* by an alternative RNA polymerase sigma factor. Proceedings of the National Academy of Sciences United States of America. 2001;**98**(10):5844-5849

[26] Mani N, Lyras D, Barroso L, Howarth P, Wilkins T, Rood JI, et al. Environmental response and autoregulation of *Clostridium difficile* TxeR, a sigma factor for toxin gene expression. Journal of Bacteriology. 2002 Nov 23;**184**(21):5971-5978

[27] El Meouche I, Peltier J, Monot M, Soutourina O, Pestel-Caron M, Dupuy B, et al. Characterization of the SigD regulon of *C. difficile* and its positive control of toxin production through the regulation of tcdR. PLoS One. 2013;**8**(12):e83748

[28] Dupuy B, Matamouros S. Regulation of toxin and bacteriocin synthesis in Clostridium species by a new subgroup of RNA polymerase σ-factors. Research in Microbiology. 2006;**157**:201-205

[29] Carter GP, Douce GR, Govind R, Howarth PM, Mackin KE, Spencer J, et al. The anti-sigma factor Tcdc modulates hypervirulence in an epidemic BI/NAP1/027 clinical isolate of *Clostridium difficile*. PLoS Pathogens. 2011;**7**(10):e1002317

[30] van Leeuwen HC, Bakker D, Steindel P, Kuijper EJ, Corver J. *Clostridium difficile* TcdC protein binds four-stranded G-quadruplex structures. Nucleic Acids Research. 2013 Feb 1;**41**(4):2382-2393

[31] Sundriyal A, Roberts AK, Ling R, Mcglashan J, Shone CC, Acharya KR. Expression, purification and cell cytotoxicity of actin-modifying binary toxin from *Clostridium difficile*. Protein Expression and Purification. 2010;**74**:42-48

[32] Schwan C, Kruppke AS, Nölke T, Schumacher L, Koch-nolte F, Kudryashev M. *Clostridium difficile* toxin CDT hijacks microtubule organization and reroutes vesicle

traffic to increase pathogen adherence. Proceedings of the National Academy of Sciences United States of America. 2014;**111**(6):2313-2318

[33] Gerding DN, Johnson S, Rupnik M, Aktories K. *Clostridium difficile* binary toxin CDT: Mechanism, epidemiology, and potential clinical importance. Gut Microbes. 2014;**5**(1):15-27

[34] Cowardin CA, Buonomo EL, Saleh MM, Wilson MG, Burgess SL, Kuehne SA, et al. The binary toxin CDT enhances *Clostridium difficile* virulence by suppressing protective colonic eosinophilia. Nature Microbiology. 2016 Jul 11;**1**:16108

[35] Sorg JA, Sonenshein AL. Bile salts and glycine as cogerminants for *Clostridium difficile* spores. Journal of Bacteriology. 2008;**190**(7):2505-2512

[36] Lawley TD, Clare S, Walker AW, Goulding D, Stabler RA, Croucher N, et al. Antibiotic treatment of *Clostridium difficile* carrier mice triggers a supershedder state, spore-mediated transmission, and severe disease in immunocompromised hosts. Infection and Immunity. 2009;**77**(9):3661-3669

[37] Deakin LJ, Clare S, Fagan RP, Dawson LF, Pickard DJ, West MR, et al. The *Clostridium difficile* spo0A gene is a persistence and transmission factor. Infection and Immunity. 2012;**80**(8):2704-2711

[38] Thapa T, Leuzzi R, Ng YK, Baban ST, Adamo R, Kuehne SA, et al. Multiple factors modulate biofilm formation by the anaerobic pathogen *Clostridium difficile*. Journal of Bacteriology. 2013;**195**(3):545-555

[39] Semenyuk EG, Laning ML, Foley J, Johnston PF, Knight KL, Gerding DN, et al. Spore formation and toxin production in *Clostridium difficile* biofilms. PLoS One. 2014;**9**(1):e87757

[40] Lawley TD, Clare S, Walker AW, Stares MD, Connor TR, Raisen C, et al. Targeted restoration of the intestinal microbiota with a simple, defined bacteriotherapy resolves relapsing *Clostridium difficile* disease in mice. PLOS Pathogens. 2012 Oct 25;**8**(10):e1002995

[41] Abecasis AB, Serrano M, Alves R, Quintais L, Pereira-Leal JB, Henriques AO. A genomic signature and the identification of new sporulation genes. Journal of Bacteriology. 2013;**195**(9):2101-2115

[42] Sarker MR, Paredes-Sabja D. Molecular basis of early stages of *Clostridium difficile* infection: Germination and colonization. Future Microbiology. 2012 Aug;**7**(8):933-943

[43] Awad MM, Johanesen PA, Carter GP, Rose E, Lyras D. *Clostridium difficile* virulence factors: Insights into an anaerobic spore-forming pathogen. Gut Microbes. 2014;**5**(5):579-593

[44] Nicholson WL, Munakata N, Horneck G, Melosh HJ, Setlow P. Resistance of *Bacillus* endospores to extreme terrestrial and extraterrestrial environments. Microbiology and Molecular Biology Reviews. 2000;**64**(3):548-572

[45] Maroo S, Lamont JT. Recurrent *Clostridium difficile*. Gastroenterology. 2006;**130**(4): 1311-1316

[46] Setlow P. Spore resistance properties. Microbiology Spectrum. 2014 Oct;**2**(5):TBS–0003-2012

[47] Heap JT, Pennington OJ, Cartman ST, Carter GP, Minton NP. The ClosTron : A universal gene knock-out system for the genus Clostridium. Journal of Microbiology Methods. 2007;**70**:452-464

[48] Ng YK, Ehsaan M, Philip S, Collery MM, Janoir C, Collignon A, et al. Expanding the repertoire of gene tools for precise manipulation of the *Clostridium difficile* genome: Allelic exchange using pyrE alleles. PLoS One. 2013;**8**(2):e56051

[49] Lawley TD, Croucher NJ, Yu L, Clare S, Sebaihia M, Goulding D, et al. Proteomic and genomic characterization of highly infectious *Clostridium difficile* 630 spores. Journal of Bacteriology. 2009;**191**(17):5377-5386

[50] Permpoonpattana P, Phetcharaburanin J, Mikelsone A, Dembek M, Tan S, Ragione R La, et al. Functional characterization of *Clostridium difficile* spore coat. Journal of Bacteriology. 2013;**195**(7):1492-1503

[51] Fimlaid KA, Bond JP, Schutz KC, Putnam EE, Leung JM, Lawley TD, et al. Global analysis of the sporulation pathway of *Clostridium difficile*. PLoS Genetics. 2013;**9**(8):e1003660

[52] Abhyankar W, Hossain AH, Djajasaputra A, Permpoonpattana P, Beek A Ter, Dekker HL, et al. In pursuit of protein targets: Proteomic characterization of bacterial spore outer layers. Journal of Proteome Research. 2013;**12**:4507-4521

[53] Pereira FC, Saujet L, Tomé AR, Serrano M, Monot M, Couture-Tosi E, et al. The spore differentiation pathway in the enteric pathogen *Clostridium difficile*. PLoS Genetics. 2013;**9**(10):e1003782

[54] Saujet L, Pereira C, Serrano M, Soutourina O, Monot M, Saujet L, et al. Genome-wide analysis of cell type-specific gene transcription during spore formation in *Clostridium difficile*. PLoS Genetics. 2013;**9**(10):e1003756

[55] Fimlaid KA, Shen A. Diverse mechanisms regulate sporulation sigma factor activity in the Firmicutes. Current Opinion in Microbiology. 2015;**24**:88-95

[56] Saujet L, Pereira C, Henriques AO, Martin-verstraete I. The regulatory network controlling spore formation in *Clostridium difficile*. FEMS Microbiology Letters. 2014;**358**:1-10

[57] Edwards AN, McBride SM. Initiation of sporulation in *Clostridium difficile*: A twist on the classic model. FEMS Microbiology Letters. 2014;**358**(2):110-118

[58] Paredes-Sabja D, Shen A, Sorg JA. *Clostridium difficile* spore biology: Sporulation, germination, and spore structural proteins. Trends in Microbiology. 2014;**22**(7):406-416

[59] Tan IS, Ramamurthi KS. Spore formation in *Bacillus subtilis*. Environmental Microbiology Reports. 2014 Jun 1;**6**(3):212-225

[60] Cassona CP, Pereira F, Serrano M, Henriques AO. A fluorescent reporter for single cell analysis of gene expression in *Clostridium difficile*. Methods in Molecular Biology. 2016;**1476**:69-90

[61] Setlow P. Mechanisms for the prevention of damage to DNA in spores of *Bacillus* species. Annual Review in Microbiology. 1995;**49**:29-54

[62] Setlow P. Spores of *Bacillus subtilis*: Their resistance to and killing by radiation, heat and chemicals. Journal of Applied Microbiology. 2006;**101**:514-525

[63] Cleveland EF, Gilvarg C. Selective degradation of peptidoglycan from *Bacillus* megaterium spores during germination. In: Gerhardt P, Costilow RN, Sadoff HL, editors. Spores VI. Washington, D.C.: American Society for Microbiology (ASM); 1978. pp. 458-464

[64] Henriques AO, Moran CPJ. Structure, assembly, and function of the spore surface layers. Annual Review in Microbiology. 2007;**61**:555-588

[65] McKenney PT, Driks A, Eichenberger P. The *Bacillus subtilis* endospore: Assembly and functions of the multilayered coat. Nature Reviews Microbiology. 2013 Jan;**11**(1):33-44

[66] Stewart GC. The exosporium layer of bacterial spores: A connection to the environment and the infected host. Microbiology and Molecular Biology Reviews. 2015;**79**(4):437-457

[67] Permpoonpattana P, Tolls EH, Nadem R, Tan S, Brisson A, Cutting SM. Surface layers of *Clostridium difficile* endospores. Journal of Bacteriology. 2011;**193**(23):6461-6470

[68] Pizarro-Guajardo M, Olguín-Araneda V, Barra-Carrasco J, Brito-Silva C, Sarker MR, Paredes-Sabja D. Characterization of the collagen-like exosporium protein, BclA1, of *Clostridium difficile* spores. Anaerobe. 2014;**25**:18-30

[69] Barra-Carrasco J, Olguín-araneda V, Plaza-garrido Á, Miranda-cárdenas C, Cofré-araneda G. The *Clostridium difficile* exosporium cysteine (CdeC)-rich protein is required for exosporium morphogenesis and coat assembly. Journal of Bacteriology. 2013;**195**(17):3863-3875

[70] Díaz-González F, Milano M, Olguin-Araneda V, Pizarro-Cerda J, Castro-Córdova P, Tzeng S-C, et al. Protein composition of the outermost exosporium-like layer of *Clostridium difficile* 630 spores. Journal of Proteomics. 2015 Jun 18;**123**:1-13

[71] Pizarro-Guajardo M, Calderón-Romero P, Castro-Córdova P, Mora-Uribe P, Paredes-Sabja D. Ultrastructural variability of the exosporium layer of *Clostridium difficile* spores. Applied and Environmental Microbiology. 2016;**82**(7):2202-2209

[72] Escobar-cortés K, Barra-carrasco J, Paredes-sabja D. Proteases and sonication specifically remove the exosporium layer of spores of *Clostridium difficile* strain 630. Journal of Microbiology Methods. 2013;**93**:25-31

[73] Phetcharaburanin J, Hong HA, Colenutt C, Bianconi I, Sempere L, Permpoonpattana P, et al. The spore-associated protein BclA1 affects the susceptibility of animals to colonization and infection by *Clostridium difficile*. Molecular Microbiology. 2014 Jun 1;**92**(5):1025-1038

[74] Janoir C, Denève C, Bouttier S, Barbut F, Hoys S, Caleechum L, et al. Adaptive strategies and pathogenesis of *Clostridium difficile* from In vivo transcriptomics. Infection and Immunity. 2013;**81**(10):3757-3769

[75] Wilson KH, Kennedy MJ, Fekety FR. Use of sodium taurocholate to enhance spore recovery on a medium selective for *Clostridium difficile*. Journal of Clinical Microbiology. 1982;**15**(3):443-446

[76] Sorg JA, Sonenshein AL. Chenodeoxycholate is an inhibitor of *Clostridium difficile* spore germination. Journal of Bacteriology. 2009;**191**(3):1115-1117

[77] Buffie CG, Bucci V, Stein RR, McKenney PT, Ling L, Gobourne A, et al. Precision microbiome restoration of bile acid-mediated resistance to *Clostridium difficile*. Nature. 2015 Jan;**517**(7533):205-208

[78] Giel JL, Sorg JA, Sonenshein AL, Zhu J. Metabolism of bile salts in mice influences spore germination in *Clostridium difficile*. PLoS One. 2010;**5**(1):e8740

[79] Wheeldon LJ, Worthington T, Lambert PA. Histidine acts as a co-germinant with glycine and taurocholate for *Clostridium difficile* spores. Journal of Applied Microbiology. 2011;**110**(4):987-994

[80] Francis MB, Allen CA, Sorg JA. Spore cortex hydrolysis precedes dipicolinic acid release during *Clostridium difficile* spore germination. Journal of Bacteriology. 2015;**197**(14):2276-2283

[81] Paidhungat M, Setlow P. Germination and outgrowth. In: Sonenshein A, Losick R, Hoch J, editors. *Bacillus subtilis* and Its Closest Relatives. Washington, D.C.: American Society for Microbiology (ASM); 2002. pp. 537-548

[82] Francis MB, Allen CA, Shrestha R, Sorg JA. Bile acid recognition by the *Clostridium difficile* germinant receptor, Cspc, is important for establishing infection. PLoS Pathogens. 2013;**9**(5):e1003356

[83] Francis MB, Sorg JA. Dipicolinic acid release by germinating *Clostridium difficile* spores occurs through a mechanosensing mechanism. mSphere. 2016;**1**(6):e00306-e00316

[84] Horsburgh MJ, Thackray PD, Moir A. transcriptional responses during outgrowth of *Bacillus subtilis* endospores. Microbiology. 2001;**147**:2933-2941

[85] Dembek M, Stabler RA, Witney AA, Wren BW, Fairweather NF. Transcriptional analysis of temporal gene expression in germinating *Clostridium difficile* 630 endospores. PLoS One. 2013;**8**(5):e64011

[86] Eckburg PB, Bik EM, Bernstein CN, Purdom E, Dethlefsen L, Sargent M, et al. Diversity of the human intestinal microbial flora. Science. 2005 Jun;**308**(5728):1635-1638

[87] Tremaroli V, Bäckhed F. Functional interactions between the gut microbiota and host metabolism. Nature. 2012;**489**(7415):242-249

[88] Nicholson JK, Holmes E, Kinross J, Burcelin R, Gibson G, Jia W, et al. Host-gut microbiota metabolic interactions. Science (80-). 2012 Jun;**336**(6086):1262 LP–1267

[89] Neish AS. Microbes in gastrointestinal health and disease. Gastroenterology. 2009;**136**(1): 65-80

[90] Carding S, Verbeke K, Vipond DT, Corfe BM, Owen LJ. Dysbiosis of the gut microbiota in disease. Microbial Ecology in Health Disease. 2015;**26**:26191

[91] Langdon A, Crook N, Dantas G. The effects of antibiotics on the microbiome throughout development and alternative approaches for therapeutic modulation. Genome Medicine. 2016 Apr;**8**:39

[92] Dethlefsen L, Relman DA. Incomplete recovery and individualized responses of the human distal gut microbiota to repeated antibiotic perturbation. Proceedings of the National Academy of Sciences United States of America. 2011 Mar;**108**(Suppl 1):4554-4561

[93] Antonopoulos DA, Huse SM, Morrison HG, Schmidt TM, Sogin ML, Young VB. Reproducible community dynamics of the gastrointestinal microbiota following antibiotic perturbation. Infection and Immunity. 2009 Jun;**77**(6):2367-2375

[94] Buffie CG, Jarchum I, Equinda M, Lipuma L, Gobourne A, Viale A, et al. Profound alterations of intestinal microbiota following a single dose of clindamycin results in sustained susceptibility to *Clostridium difficile*-induced colitis. McCormick BA, editor. Infection and Immunity. 2012 Jan;**80**(1):62-73

[95] Jernberg C, Lofmark S, Edlund C, Jansson JK. Long-term ecological impacts of antibiotic administration on the human intestinal microbiota. ISME Journal. 2007; **1**(1):56-66

[96] Pérez-Cobas AE, Gosalbes MJ, Friedrichs A, Knecht H, Artacho A, Eismann K, et al. Gut microbiota disturbance during antibiotic therapy: A multi-omic approach. Gut. 2013 Nov;**62**(11):1591-1601

[97] Bassis CM, Theriot CM, Young VB. Alteration of the murine gastrointestinal microbiota by tigecycline leads to increased susceptibility to clostridium difficile infection. Antimicrobial Agents and Chemotherapy. 2014;**58**(5):2767-2774

[98] Theriot CM, Koenigsknecht MJ, Carlson PE, Hatton GE, Nelson AM, Li B, et al. Antibiotic-induced shifts in the mouse gut microbiome and metabolome increase susceptibility to *Clostridium difficile* infection. Nature Communication. 2014;**5**:3114

[99] Zhao Y, Wu J, Li J V, Zhou N-Y, Tang H, Wang Y. Gut microbiota composition modifies fecal metabolic profiles in mice. Journal of Proteome Research. 2013 Jun;**12**(6):2987-2999

[100] Scaria J, Janvilisri T, Fubini S, Gleed RD, McDonough SP, Chang Y-F. *Clostridium difficile* transcriptome analysis using pig ligated loop model reveals modulation of pathways not modulated in vitro. Journal of Infectious Diseases. 2011 Jun;**203**(11):1613-1620

[101] Furusawa Y, Obata Y, Fukuda S, Endo TA, Nakato G, Takahashi D, et al. Commensal microbe-derived butyrate induces the differentiation of colonic regulatory T cells. Nature. 2013 Dec 19;**504**(7480):446-450

[102] Wong JMW, de Souza R, Kendall CWC, Emam A, Jenkins DJA. Colonic health: Fermentation and short chain fatty acids. Journal of Clinical Gastroenterology. 2006;**40**(3): 235-243

[103] Martin-Verstraete I, Peltier J, Dupuy B. The regulatory networks that control *Clostridium difficile* toxin synthesis. Toxins (Basel). 2016 May;**8**(5):153

[104] Darkoh C, DuPont HL, Norris SJ, Kaplan HB. Toxin synthesis by *Clostridium difficile* is regulated through quorum signaling. MBio. 2015;**6**(2):e02569–e025614

[105] Darkoh C, Odo C, DuPont HL. Accessory gene regulator-1 locus is essential for virulence and pathogenesis of *Clostridium difficile*. MBio. 2016 Sep 7;**7**(4):e01237–e012316

[106] Martin MJ, Clare S, Goulding D, Faulds-Pain A, Barquist L, Browne HP, et al. The agr locus regulates virulence and colonization genes in *Clostridium difficile* 027. Journal of Bacteriology. 2013;**195**(16):3672-3681

[107] Aubry A, Hussack G, Chen W, KuoLee R, Twine SM, Fulton KM, et al. Modulation of toxin production by the flagellar regulon in *Clostridium difficile*. Infection and Immunity. 2012 Oct;**80**(10):3521-3532

[108] Bouillaut L, Dubois T, Sonenshein AL, Dupuy B. Integration of metabolism and virulence in *Clostridium difficile*. Research in Microbiology. 2015 May;**166**(4):375-383

[109] Pettit LJ, Browne HP, Yu L, Smits WK, Fagan RP, Barquist L, et al. Functional genomics reveals that *Clostridium difficile*Spo0A coordinates sporulation, virulence and metabolism. BMC Genomics. 2014;**15**(1):160

[110] Purcell EB, McKee RW, McBride SM, Waters CM, Tamayo R. Cyclic diguanylate inversely regulates motility and aggregation in *Clostridium difficile*. Journal of Bacteriology. 2012 Jul;**194**(13):3307-3316

[111] McKee RW, Mangalea MR, Purcell EB, Borchardt EK, Tamayo R. The second messenger cyclic Di-GMP regulates *Clostridium difficile* toxin production by controlling expression of sigD. Journal of Bacteriology. 2013 Nov;**195**(22):5174-5185

[112] Edwards AN, Tamayo R, McBride SM. A novel regulator controls *Clostridium difficile* sporulation, motility and toxin production. Molecular Microbiology. 2016 Jun;**100**(6):954-971

[113] Schleifer KH, Kandler O. Peptidoglycan types of bacterial cell walls and their taxonomic implications. Bacteriological Reviews. 1972;**36**(4):407-477

[114] Vollmer W, Blanot D, De Pedro MA. Peptidoglycan structure and architecture. FEMS Microbiology Reviews. 2008;**32**(2):149-167

[115] Kirk JA, Banerji O, Fagan RP. Characteristics of the *Clostridium difficile* cell envelope and its importance in therapeutics. Microbial Biotechnology. 2017;10(1):76-90

[116] Turner RD, Vollmer W, Foster SJ. Different walls for rods and balls: The diversity of peptidoglycan. Molecular Microbiology. 2014;**91**(5):862-874

[117] Macheboeuf P, Contreras-Martel C, Job V, Dideberg O, Dessen A. Penicillin binding proteins: Key players in bacterial cell cycle and drug resistance processes. FEMS Microbiology Reviews. 2006;**30**(5):673-691

[118] Bugg TDH, Braddick D, Dowson CG, Roper DI. Bacterial cell wall assembly: Still an attractive antibacterial target. Trends in Biotechnology. 2011;**29**(4):167-173

[119] Smith CA. Structure, function and dynamics in the mur family of bacterial cell wall ligases. Journal of Molecular Biology. 2006;**362**(4):640-655

[120] Barreteau H, Kovac A, Boniface A, Sova M, Gobec S, Blanot D. Cytoplasmic steps of peptidoglycan biosynthesis. FEMS Microbiology Reviews. 2008 Mar;**32**(2):168-207

[121] Typas A, Banzhaf M, Gross CA, Vollmer W. From the regulation of peptidoglycan synthesis to bacterial growth and morphology. Nature Reviews Microbiology. 2012;**10**(2):123-136

[122] Leeds JA, Sachdeva M, Mullin S, Whitney Barnes S, Ruzin A. *In vitro* selection, via serial passage, of *Clostridium difficile* mutants with reduced susceptibility to fidaxomicin or vancomycin. Journal of Antimicrobial Chemotherapy. 2014;**69**(1):41-44

[123] Meyer P, Gutierrez J, Pogliano K, Dworkin J. Cell wall synthesis is necessary for membrane dynamics during sporulation of *Bacillus subtilis*. Molecular Microbiology. 2010;**76**(4):956-970

[124] Bouhss A, Trunkfield AE, Bugg TDH, Mengin-Lecreulx D. The biosynthesis of peptidoglycan lipid-linked intermediates. FEMS Microbiology Reviews. 2008;**32**(2):208-233

[125] Fay A, Meyer P, Dworkin J. Interactions between late-acting proteins required for peptidoglycan synthesis during sporulation. Journal of Molecular Biology. 2010 Jun 18;**399**(4):547-561

[126] Vasudevan P, McElligott J, Attkisson C, Betteken M, Popham DL. Homologues of the *Bacillus subtilis* SpoVB protein are involved in cell wall metabolism. Journal of Bacteriology. 2009;**191**(19):6012-6019

[127] Meeske AJ, Sham L-T, Kimsey H, Koo B-M, Gross C a, Bernhardt TG, et al. MurJ and a novel lipid II flippase are required for cell wall biogenesis in *Bacillus subtilis*. Proceedings of the National Academy of Sciences United States of America. 2015;**112**(20):6437-6442

[128] Scheffers D, Pinho M. Bacterial cell wall synthesis: New insights from localization studies. Microbiology and Molecular Biology Reviews. 2005;**69**(4):585-607

[129] Packiam M, Weinrick B, Jacobs WR, Maurelli AT. Structural characterization of muropeptides from Chlamydia trachomatis peptidoglycan by mass spectrometry resolves "chlamydial anomaly". Proceedings of the National Academy of Sciences United States of America. 2015;**112**(37):11660-11665

[130] Mcpherson DC, Popham DL. Peptidoglycan synthesis in the absence of Class A penicillin-binding proteins in *Bacillus subtilis*. Journal of Bacteriology. 2003;**185**(4):1423-1431

[131] Arbeloa A, Segal H, Hugonnet J-E, Josseaume N, Dubost L, Brouard J-P, et al. Role of class A penicillin-binding proteins in PBP5-mediated beta-lactam resistance in *Enterococcus faecalis*. Journal of Bacteriology. 2004;**186**(5):1221-1228

[132] Rice LB, Carias LL, Rudin S, Hutton R, Marshall S, Hassan M, et al. Role of class A penicillin-binding proteins in the expression of β-lactam resistance in *Enterococcus faecium*. Journal of Bacteriology. 2009;**191**(11):3649-3656

[133] Henriques AO, Glaser P, Piggot PJ, Moran Jr CP. Control of cell shape and elongation by the rodA gene in *Bacillus subtilis*. Molecular Microbiology. 1998 Apr 1;**28**(2):235-247

[134] Meeske AJ, Riley EP, Robins WP, Uehara T, Mekalanos JJ, Kahne D, et al. SEDS proteins are a widespread family of bacterial cell wall polymerases. Nature. 2016 Sep;**537**(7622):634-638

[135] Cho H, Wivagg CN, Kapoor M, Barry Z, Rohs PDA, Suh H, et al. Bacterial cell wall biogenesis is mediated by SEDS and PBP polymerase families functioning semi-autonomously. Nature Microbiology. 2016 Sep 19;**1**:16172

[136] Emami K, Guyet A, Kawai Y, Devi J, Wu LJ, Allenby N, et al. RodA as the missing glycosyltransferase in *Bacillus subtilis* and antibiotic discovery for the peptidoglycan polymerase pathway. Nature Microbiology. 2017 January;**2**:16253

[137] Real G, Fay A, Eldar A, Pinto SM, Henriques AO, Dworkin J. Determinants for the subcellular localization and function of a nonessential SEDS protein. Journal of Bacteriology. 2008;**190**(1):363-376

[138] Peltier J, Courtin P, El Meouche I, Lemée L, Chapot-Chartier MP, Pons JL. *Clostridium difficile* has an original peptidoglycan structure with a high level of N-acetylglucosamine deacetylation and mainly 3-3 cross-links. Journal of Biological Chemistry. 2011;**286**(33):29053-29062

[139] Ho TD, Williams KB, Chen Y, Helm RF, Popham DL, Ellermeier CD. *Clostridium difficile* extracytoplasmic function σ factor σV regulates lysozyme resistance and is necessary for pathogenesis in the hamster model of infection. Infection and Immunity. 2014;**82**(6):2345-2355

[140] Vollmer W, Tomasz A. The pgdA gene encodes for a peptidoglycan N-acetylglucosamine deacetylase in *Streptococcus pneumoniae*. Journal of Biological Chemistry. 2000;**275**(27):20496-20501

[141] Boneca IG, Dussurget O, Cabanes D, Nahori M-A, Sousa S, Lecuit M, et al. A critical role for peptidoglycan N-deacetylation in Listeria evasion from the host innate immune system. Proceedings of the National Academy of Sciences United States of America. 2007;**104**(3):997-1002

[142] Mainardi J-L, Villet R, Bugg TD, Mayer C, Arthur M. Evolution of peptidoglycan biosynthesis under the selective pressure of antibiotics in Gram-positive bacteria. FEMS Microbiology Reviews. 2008 Mar 1;**32**(2):386-408

[143] Ammam F, Meziane-cherif D, Mengin-Lecreulx D, Blanot D, Patin D, Boneca IG, et al. The functional vanGCd cluster of *Clostridium difficile* does not confer vancomycin resistance. Molecular Microbiology. 2013;**89**(4):612-625

[144] Ghosh AS, Chowdhury C, Nelson DE. Physiological functions of D-alanine carboxypeptidases in *Escherichia coli*. Trends Microbiol. 2008;**16**(7):309-17.

[145] McBride SM, Sonenshein AL. The dlt operon confers resistance to cationic antimicrobial peptides in Clostridium difficile. Microbiology. 2011 May 1;**157**(5):1457-65.

[146] Freeman J, Vernon J, Morris K, Nicholson S, Todhunter S, Longshaw C, et al. Pan-European longitudinal surveillance of antibiotic resistance among prevalent *Clostridium difficile* ribotypes. Clinical Microbiology Infection. 2015;**21**(3):248.e9-248.e16

[147] Tickler IA, Goering RV, Whitmore JD, Lynn ANW, Persing DH, Tenover FC. Strain types and antimicrobial resistance patterns of *Clostridium difficile* isolates from the United States, 2011 to 2013. Antimicrobial Agents and Chemotherapy. 2014;**58**(7):4214-4218

[148] Zapun A, Contreras-Martel C, Vernet T. Penicillin-binding proteins and β-lactam resistance. FEMS Microbiology Reviews. 2008;**32**(2):361-385

[149] Pazos M, Peters K, Vollmer W. Robust peptidoglycan growth by dynamic and variable multi-protein complexes. Current Opinion in Microbiology. 2017 Feb;**36**:55-61

[150] Sauvage E, Kerff F, Terrak M, Ayala JA, Charlier P. The penicillin-binding proteins: Structure and role in peptidoglycan biosynthesis. FEMS Microbiology Reviews. 2008;**32**(2):234-258

[151] Zapun A, Vernet T, Pinho MG. The different shapes of *cocci*. FEMS Microbiology Reviews. 2008;**32**(2):345-360

[152] Isidro J, Santos A, Nunes A, Borges V, Silva C, Chaves C, et al. Imipenem resistance in *Clostridium difficile* epidemic ribotype 017, Portugal. Emerging Infectious Diseases. 2017 (Submitted)

[153] He M, Sebaihia M, Lawley TD, Stabler RA, Dawson LF, Martin MJ, et al. Evolutionary dynamics of *Clostridium difficile* over short and long time scales. Proceedings of the National Academy of Sciences United States of America. 2010;**107**(16):7527-7532

[154] Yang DC, Blair KM, Salama NR. Staying in shape: The impact of cell shape on bacterial survival in diverse environments. Microbiology and Molecular Biology Reviews. 2016 Mar;**80**(1):187-203

[155] Matteï PJ, Neves D, Dessen A. Bridging cell wall biosynthesis and bacterial morphogenesis. Current Opinion in Structural Biology. 2010;**20**(6):749-755

[156] Stewart GC. Taking shape: Control of bacterial cell wall biosynthesis. Molecular Microbiology. 2005;**57**(5):1177-1181

[157] Egan AJF, Cleverley RM, Peters K, Lewis RJ, Vollmer W. Regulation of bacterial cell wall growth. FEBS Journal. 2017;**284**(6):851-867

[158] Typas A, Banzhaf M, Gross CA, Vollmer W. From the regulation of peptidoglycan synthesis to bacterial growth and morphology. Nature Review Microbiology. 2011 Dec;**10**(2):123-136

[159] Egan AJF, Biboy J, van't Veer I, Breukink E, Vollmer W. Activities and regulation of peptidoglycan synthases. Philosophical Transactions of the Royal Society of Lond B: Biological Science. 2015 Oct;**370**(1679)

[160] Den Blaauwen T, De Pedro MA, Nguyen-Distèche M, Ayala JA. Morphogenesis of rod-shaped sacculi. FEMS Microbiology Reviews. 2008;**32**(2):321-344

[161] Haeusser DP, Margolin W. Splitsville: Structural and functional insights into the dynamic bacterial Z ring. Nature Review Microbiology. 2016 Apr;**14**(5):305-319

[162] Errington J, Daniel RA, Scheffers D-J. Cytokinesis in Bacteria. Microbiology and Molecular Biology Review. 2003;**67**(1):52-65

[163] Haydon DJ, Stokes NR, Ure R, Galbraith G, Bennett JM, Brown DR, et al. An inhibitor of FtsZ with potent and selective anti-Staphylococcal activity. Science (80-). 2008 Sep 19;**321**(5896):1673 LP-1675

[164] Stokes NR, Baker N, Bennett JM, Berry J, Collins I, Czaplewski LG, et al. An improved small-molecule inhibitor of FtsZ with superior in vitro potency, drug-like properties, and in vivo efficacy. Antimicrobial Agents and Chemotherapy. 2013;**57**(1):317-325

[165] Sun N, Chan F-Y, Lu Y-J, Neves MAC, Lui H-K, Wang Y, et al. Rational design of berberine-based FtsZ inhibitors with broad-spectrum antibacterial activity. PLoS One. 2014 May 13;**9**(5):e97514

[166] Ransom EM, Williams KB, Weiss DS, Ellermeier CD. Identification and characterization of a gene cluster required for proper rod shape, cell division, and pathogenesis in *Clostridium difficile*. Journal of Bacteriology. 2014;**196**(12):2290-2300

[167] Bukowska-Faniband E, Hederstedt L. The PASTA domain of penicillin-binding protein SpoVD is dispensable for endospore cortex peptidoglycan assembly in *Bacillus subtilis*. Microbiology. 2015;**161**(2015):330-340

[168] Bukowska-Faniband E, Hederstedt L. Cortex synthesis during *Bacillus subtilis* sporulation depends on the transpeptidase activity of SpoVD. FEMS Microbiology Letters. 2013;**346**(1):65-72

[169] Dancer BN. Requirement for peptidoglycan synthesis during sporulation of *Bacillus subtilis*. Journal of Bacteriology. 1979;**140**(3):786-797

[170] Ojkic N, López-Garrido J, Pogliano K, Endres RG. Cell-wall remodeling drives engulfment during *Bacillus subtilis* sporulation. In: Elife. Elife. 2016;**5**:e18657

[171] Popham DL, Gilmore ME, Setlow P, Documentacion SDE. Roles of low-molecular-weight penicillin-binding proteins in *Bacillus subtilis* spore peptidoglycan synthesis and spore properties. Journal of Bacteriology. 1999;**181**(1):126-132

[172] Orsburn B, Sucre K, Popham DL, Melville SB. The SpmA/B and DacF proteins of *Clostridium perfringens* play important roles in spore heat resistance: Research letter. FEMS Microbiology Letters. 2009;**291**(2):188-194

[173] Gilmore ME, Bandyopadhyay D, Dean AM, Linnstaedt SD, Popham DL. Production of muramic δ-lactam in *Bacillus subtilis* spore peptidoglycan. Journal of Bacteriology. 2004;**186**(1):80-89

[174] Popham DL, Helin J, Costello CE, Setlow P. Muramic lactam in peptidoglycan of *Bacillus subtilis* spores is required for spore outgrowth but not for spore dehydration or heat resistance. Proceedings of the National Academy of Sciences. 1996;**93**(26):15405-15410

[175] Brown KA, Khanafer N, Daneman N, Fisman DN. Meta-analysis of antibiotics and the risk of community-associated *Clostridium difficile* infection. Antimicrobial Agents and Chemotherapy. 2013;**57**(5):2326-2332

[176] Deshpande A, Pasupuleti V, Thota P, Pant C, Rolston DDK, Sferra TJ, et al. Community-associated clostridium difficile infection antibiotics: A meta-analysis. Journal of Antimicrobial Chemotherapy. 2013;**68**(9):1951-1961

[177] Slimings C, Riley TV. Antibiotics and hospital-acquired *Clostridium difficile* infection: Update of systematic review and meta-analysis. Journal of Antimicrobial Chemotherapy. 2014;**69**(4):881-891

[178] Vardakas KZ, Konstantelias AA, Loizidis G, Rafailidis PI, Falagas ME. Risk factors for development of *Clostridium difficile* infection due to BI/NAP1/027 strain: A meta-analysis. International Journal of Infectious Diseases. 2012;**16**(11):0-5

[179] Pépin J, Saheb N, Coulombe M-A, Alary M-E, Marie-Pier Corriveau SA. Emergence of fluoroquinolones as the predominant risk factor for Clostridium difficile-associated diarrhea: a cohort study during an epidemic in Quebec. Clinical Infectious Diseases. 2005;**41**(9):1254-1260

[180] Kourbeti IS, Alegakis DE, Maraki S, Samonis G. Impact of prolonged treatment with high-dose ciprofloxacin on human gut flora: A case report. Journal of Medical Case Reports. 2010;**4**(1):111

[181] Gerding DN. Clindamycin, cephalosporins, fluoroquinolones, and *Clostridium difficile*–associated diarrhea: This is an antimicrobial resistance problem. Clinical Infectious Diseases. 2004;**38**(5):646-648

[182] Johnson S, Samore MH, Farrow KA, Killgore GE, Tenover FC, Lyras D, et al. Epidemics of diarrhea caused by a clindamycin-resistant strain of *Clostridium difficile* in four hospitals. New England Journal of Medicine. 1999 Nov;**341**(22):1645-1651

[183] O'Connor JR, Johnson S, Gerding DN. *Clostridium difficile* infection caused by the epidemic BI/NAP1/027 strain. Gastroenterology. 2009;**136**(6):1913-1924

[184] He M, Miyajima F, Roberts P, Ellison L, Pickard DJ, Martin MJ, et al. Emergence and global spread of epidemic healthcare-associated *Clostridium difficile*. Nature Genetics. 2013;**45**(1):109-113

[185] Kuijper EJ, Barbut F, Brazier JS, Kleinkauf N, Eckmanns T, Lambert ML, Drudy D, Fitzpatrick F, Wiuff C, Brown DJ, Coia JE, Pituch H, Reichert P, Even J, Mossong J,

Widmer AF, Olsen KE, Allerberger F, Notermans DW, Delmée M, Coignard B, Wilcox M, Patel B, MD. Update of Clostridium difficile infection due to PCR ribotype 027 in Europe, 2008. Eurosurveillance. 2008:**13**(31):pii=18942

[186] Linder JA, Huang ES, Steinman MA, Gonzales R, Stafford RS. Fluoroquinolone prescribing in the United States: 1995 to 2002. American Journal of Medicine. 2005;**118**(3):259-268

[187] Bauer MP, Notermans DW, van Benthem BH, Brazier JS, Wilcox MH, Rupnik M, et al. *Clostridium difficile* infection in Europe: A hospital-based survey. Lancet. 2011;**377**(9759):63-73

[188] Davies KA, Ashwin H, Longshaw CM, Burns DA, Davis GL, Wilcox MH. Diversity of *Clostridium difficile* PCR ribotypes in Europe: Results from the European, multicentre, prospective, biannual, point-prevalence study of *Clostridium difficile* infection in hospitalised patients with diarrhoea (EUCLID), 2012 and 2013. Eurosurveillance. 2016;**21**(29):1-11

[189] Pituch H, Obuch-Woszczatyński P, Wultańska D, Nurzyńska G, Harmanus C, Banaszkiewicz A, et al. Characterization and antimicrobial susceptibility of *Clostridium difficile* strains isolated from adult patients with diarrhoea hospitalized in two university hospitals in Poland, 2004-2006. Journal of Medical Microbiology. 2011;**60**(8):1200-1205

[190] Dobreva EG, Ivanov IN, Vathcheva-Dobrevska RS, Ivanova KI, Asseva GD, Petrov PK, et al. Advances in molecular surveillance of *Clostridium difficile* in Bulgaria. Journal of Medical Microbiology. 2013;**62**(2013):1428-1434

[191] Gao Q, Wu S, Huang H, Ni Y, Chen Y, Hu Y, et al. Toxin profiles, PCR ribotypes and resistance patterns of *Clostridium difficile*: A multicentre study in China, 2012-2013. International Journal of Antimicrobial Agents. 2016;**48**(6):736-739

[192] Lee JH, Lee Y, Lee K, Riley TV, Kim H. The changes of PCR ribotype and antimicrobial resistance of *Clostridium difficile* in a tertiary care hospital over 10 years. Journal of Medical Microbiology. 2014;**63**:819-823

[193] Putsathit P, Maneerattanaporn M, Piewngam P, Kiratisin P, Riley TV. Prevalence and molecular epidemiology of *Clostridium difficile* infection in Thailand. New Microbes and New Infections. 2017;**15**:27-32

[194] Santos A, Isidro J, Silva C, Boaventura L, Diogo J, Faustino A, et al. Molecular and epidemiologic study of *Clostridium difficile* reveals unusual heterogeneity in clinical strains circulating in different regions in Portugal. Clinical Microbiology Infections. 2016;**22**(8):695-700

[195] Spigaglia P. Recent advances in the understanding of antibiotic resistance in *Clostridium difficile* infection. Therapeutic Advances in Infectious Diseases. 2016;**3**(1):23-42

[196] Freeman J, Bauer MP, Baines SD, Corver J, Fawley WN, Goorhuis B, et al. The changing epidemiology of *Clostridium difficile* infections. Clinical Microbiology Reviews. 2010;**23**(3):529-549

[197] Blair JMA, Webber MA, Baylay AJ, Ogbolu DO, Piddock LJ V. Molecular mechanisms of antibiotic resistance. Nature Review Microbiology. 2015;**13**(1):42-51

[198] Debast SB, Bauer MP, Kuijper EJ, Diseases ES of CM and I. European Society of Clinical Microbiology and Infectious Diseases: Update of the treatment guidance document for *Clostridium difficile* infection. Clinical Microbiology Infections. 2014 March;**20**(Suppl 2):1-26

[199] Snydman DR, McDermott LA, Jacobus NV, Thorpe C, Stone S, Jenkins SG, et al. U.S.-Based National Sentinel Surveillance Study for the epidemiology of *Clostridium difficile*-associated diarrheal isolates and their susceptibility to fidaxomicin. Antimicrobial Agents and Chemotherapy. 2015;**59**(10):6437-6443

[200] Baines SD, O'Connor R, Freeman J, Fawley WN, Harmanus C, Mastrantonio P, et al. Emergence of reduced susceptibility to metronidazole in *Clostridium difficile*. Journal of Antimicrobial Chemotherapy. 2008;**62**(5):1046-1052

[201] Peláez T, Cercenado E, Alcalá L, Marín M, Martín-López A, Martínez-Alarcón J, et al. Metronidazole resistance in *Clostridium difficile* is heterogeneous. Journal of Clinical Microbiology. 2008;**46**(9):3028-3032

[202] Kuijper EJ, Wilcox MH. *Editorial Commentary:* Decreased effectiveness of metronidazole for the treatment of *Clostridium difficile* Infection? Clinical Infectious Diseases. 2008;**47**(1):63-65

[203] Richardson C, Kim P, Lee C, Bersenas A, Weese JS. Comparison of *Clostridium difficile* isolates from individuals with recurrent and single episode of infection. Anaerobe. 2015;**33**:105-108

[204] Land KM, Johnson PJ. Molecular basis of metronidazole resistance in pathogenic bacteria and protozoa. Drug Resistances Updates. 1999 Oct;**2**(5):289-294

[205] Löfmark S, Edlund C, Nord CE. Metronidazole is still the drug of choice for treatment of anaerobic infections. Clinical Infectious Diseases. 2010 Jan;**50**(s1):S16–S23

[206] Lynch T, Chong P, Zhang J, Hizon R, Du T, Graham MR, et al. Characterization of a stable, metronidazole-resistant *Clostridium difficile* clinical isolate. PLoS One. 2013;**8**(1):e53757

[207] Chong PM, Lynch T, McCorrister S, Kibsey P, Miller M, Gravel D, et al. Proteomic analysis of a NAP1 *Clostridium difficile* clinical isolate resistant to metronidazole. PLoS One. 2014;**9**(1):e82622

[208] Ernst FD, Homuth G, Stoof J, Mader U, Waidner B, Kuipers EJ, et al. Iron-responsive regulation of the *Helicobacter pylori* iron-cofactored superoxide dismutase SodB is mediated by Fur. Journal of Bacteriology. 2005 Jun 1;**187**(11):3687-3692

[209] Choi SS, Chivers PT, Berg DE. Point mutations in *Helicobacter pylori*'s fur regulatory gene that alter resistance to metronidazole, a prodrug activated by chemical reduction. PLoS One. 2011 Mar 25;**6**(3):e18236

[210] Tsugawa H, Suzuki H, Satoh K, Hirata K, Matsuzaki J, Saito Y, et al. Two amino acids mutation of ferric uptake regulator determines *Helicobacter pylori* resistance to metronidazole. Antioxidants & Redox Signaling. 2011 Jan;**14**(1):15-23

[211] Moura I, Monot M, Tani C, Spigaglia P, Barbanti F, Norais N, et al. Multidisciplinary analysis of a nontoxigenic *Clostridium difficile* strain with stable resistance to metronidazole. Antimicrobial Agents and Chemotherapy. 2014;**58**(8):4957-4960

[212] Artsimovitch I, Seddon J, Sears P. Fidaxomicin is an inhibitor of the initiation of bacterial RNA synthesis. Clinical Infectious Diseases. 2012 Aug 1;**55**(Suppl 2):S127–S131

[213] Babakhani F, Gomez A, Robert N, Sears P. Killing kinetics of fidaxomicin and its major metabolite, OP-1118, against *Clostridium difficile*. Journal of Medical Microbiology. 2011 Aug 1;**60**(8):1213-1217

[214] Seddon J, Mondal K, Sevostyanova A, Artsimovitch I. Transcription inhibition by fidaxomicin [poster]. In: 2011 FASEB Summer Research Conferences: Mechanism & Regulation of Prokaryotic Transcription; 19-24 June 2011; Saxton River, VT

[215] Goldstein EJC, Citron DM, Sears P, Babakhani F, Sambol SP, Gerding DN. Comparative susceptibilities to fidaxomicin (OPT-80) of isolates collected at baseline, recurrence, and failure from patients in two phase III trials of fidaxomicin against *Clostridium difficile* infection. Antimicrobial Agents and Chemotherapy. 2011;**55**(11):5194-5199

[216] Knight DR, Giglio S, Huntington PG, Korman TM, Kotsanas D, Moore CV, et al. Surveillance for antimicrobial resistance in Australian isolates of *Clostridium difficile*, 2013-14. Journal of Antimicrobial Chemotherapy. 2015;**70**(11):2992-2999

[217] Lachowicz D, Pituch H, Obuch-Woszczatyński P. Antimicrobial susceptibility patterns of *Clostridium difficile* strains belonging to different polymerase chain reaction ribotypes isolated in Poland in 2012. Anaerobe. 2015;**31**:37-41

[218] Spigaglia P, Barbanti F, Mastrantonio P, Ackermann G, Balmelli C, Barbut F, et al. Multidrug resistance in European *Clostridium difficile* clinical isolates. Journal of Antimicrobial Chemotherapy. 2011;**66**(10):2227-2234

[219] Norén T, Alriksson I, Åkerlund T, Burman LG, Unemo M. In vitro susceptibility to 17 antimicrobials of clinical *Clostridium difficile* isolates collected in 1993-2007 in Sweden. Clinical Microbiology Infection. 2010;**16**(8):1104-1110

[220] Knight DR, Elliott B, Chang BJ, Perkins TT, Riley T V. Diversity and evolution in the genome of *Clostridium difficile*. Clinical Microbiology Reviews. 2015;**28**(3):721-741

[221] Hansen LH, Vester B. A cfr-like gene from clostridium difficile confers multiple antibiotic resistance by the same mechanism as the cfr gene. Antimicrobial Agents and Chemotherapy. 2015;**59**(9):5841-5843

[222] Marín M, Martín A, Alcalá L, Cercenado E, Iglesias C, Reigadas E, et al. *Clostridium difficile* isolates with high linezolid MICs harbor the multiresistance gene cfr. Antimicrobial Agents and Chemotherapy. 2015;**59**(1):586-589

[223] Hooper DC. Mechanisms of action of antimicrobials: Focus on fluoroquinolones. Clinical Infectious Diseases. 2001 Mar 15;**32**(Suppl 1):S9–S15

[224] Pépin J, Saheb N, Coulombe M-A, Alary M-E, Corriveau M-P, Authier S, et al. Emergence of fluoroquinolones as the predominant risk factor for *Clostridium difficile*-associated diarrhea: A cohort study during an epidemic in Quebec. Clinical Infectious Diseases. 2005;**41**(9):1254-1260

[225] Tenover FC, Tickler IA, Persing DH. Antimicrobial-resistant strains of *Clostridium difficile* from North America. Antimicrobial Agents and Chemotherapy. 2012;**56**(6):2929-2932

[226] Huang H, Weintraub A, Fang H, Nord CE. Antimicrobial resistance in *Clostridium difficile*. International Journal of Antimicrobial Agents. 2009;**34**(6):516-522

[227] Wasels F, Kuehne SA, Cartman ST, Spigaglia P, Barbanti F, Minton NP. Fluoroquinolone resistance does not impose a cost on the fitness of *Clostridium difficile* in vitro. Antimicrobial Agents and Chemotherapy. 2015;**59**(3):1794-1796

[228] O'Connor JR, Galang MA, Sambol SP, Hecht DW, Vedantam G, Gerding DN, et al. Rifampin and rifaximin resistance in clinical isolates of *Clostridium difficile*. Antimicrobial Agents and Chemotherapy. 2008;**52**(8):2813-2817

[229] Gerding DN, Muto CA, Owens, Jr. RC. Treatment of *Clostridium difficile* infection. Clinical Infectious Diseases. 2008 Jan 15;46(s1):S32–S42

[230] Cohen SH, Gerding DN, Johnson S, Kelly CP, Loo VG, McDonald LC, et al. Clinical practice guidelines for *Clostridium difficile* infection in adults: 2010 update by the Society for Healthcare Epidemiology of America (SHEA) and the Infectious Diseases Society of America (IDSA). Infection Control and Hospital Epidemiology. 2010;**31**(5):431-455

[231] Curry SR, Marsh JW, Shutt KA, Muto CA, O'Leary MM, Saul MI, et al. High frequency of rifampin resistance identified in an epidemic *Clostridium difficile* clone from a large teaching hospital. Clinical Infectious Diseases. 2009;**48**(4):425-429

[232] Stevenson EC, Major GA, Spiller RC, Kuehne SA, Minton NP. Coinfection and emergence of rifamycin resistance during a recurrent clostridium difficile infection. Journal of Clinical Microbiology. 2016;**54**(11):2689-2694

[233] Chopra I, Roberts M. Tetracycline antibiotics: Mode of action, applications, molecular biology, and epidemiology of bacterial resistance. Microbiology and Molecular Biology Reviews. 2001 Jun 1;**65**(2):232-260

[234] Dong D, Zhang L, Chen X, Jiang C, Yu B, Wang X, et al. Antimicrobial susceptibility and resistance mechanisms of clinical *Clostridium difficile* from a Chinese tertiary hospital. International Journal of Antimicrobial Agents. 2013;**41**:80-84

[235] Barbut F, Mastrantonio P, Delmée M, Brazier J, Kuijper E, Poxton I, et al. Prospective study of *Clostridium difficile* infections in Europe with phenotypic and genotypic characterisation of the isolates. Clinical Microbiology Infections. 2007;**13**(11):1048-1057

[236] Dong D, Chen X, Jiang C, Zhang L, Cai G, Han L, et al. Genetic analysis of Tn916-like elements conferring tetracycline resistance in clinical isolates of *Clostridium difficile*. International Journal of Antimicrobial Agents. 2014;**43**(1):73-77

[237] Donhofer A, Franckenberg S, Wickles S, Berninghausen O, Beckmann R, Wilson DN. Structural basis for TetM-mediated tetracycline resistance. Proceedings of the National Academy of Sciences. 2012 Oct 16;**109**(42):16900-16905

[238] Fry PR, Thakur S, Abley M, Gebreyes WA. Antimicrobial resistance, toxinotype, and genotypic profiling of *Clostridium difficile* isolates of swine origin. Journal of Clinical Microbiology. 2012;**50**(7):2366-2372

[239] Spigaglia P, Barbanti F, Mastrantonio P. Detection of a genetic linkage between genes coding for resistance to tetracycline and erythromycin in *Clostridium difficile*. Microbial Drug Resistance. 2007;**13**(2):90-95

[240] European Surveillance of Veterinary Antimicrobial Consumption 2016. Sales of veterinary antimicrobial agents in 29 European countries in 2014 (EMA/61769/2016). European Medicines Agency

[241] Knetsch C, Connor T, Mutreja A, van Dorp S, Sanders I, Browne H, et al. Whole genome sequencing reveals potential spread of *Clostridium difficile* between humans and farm animals in the Netherlands, 2002 to 2011. Eurosurveillance. 2014 Nov 13;19(45):pii=20954

[242] Bulkley D, Innis CA, Blaha G, Steitz TA. Revisiting the structures of several antibiotics bound to the bacterial ribosome. Proceedings of the National Academy of Sciences. 2010 Oct 5;**107**(40):17158-17163

[243] Leslie AG, Moody PC, Shaw WV. Structure of chloramphenicol acetyltransferase at 1.75-A resolution. Proceedings of the National Academy of Sciences. 1988 Jun 1;**85**(12):4133-4137

[244] Lyras D, Storie C, Huggins AS, Crellin PK, Bannam TL, Rood JI. Chloramphenicol resistance in *Clostridium difficile* is encoded on Tn 4453 transposons that are closely related to Tn 4451 from Clostridium perfringens. Antimicrobial Agents and Chemotherapy. 1998;**42**(7):1563-1567

[245] European Centre for Disease Prevention and Control (ECDC). Surveillance of antimicrobial consumption in Europe 2012. Stockholm: ECDC; 2014

[246] Fawley WN, Underwood S, Freeman J, Baines SD, Saxton K, Stephenson K, et al. Efficacy of hospital cleaning agents and germicides against epidemic *Clostridium difficile* strains. Infection Control Hospital Epidemiology. 2007;**28**(8):920-925

[247] Nerandzic MM, Donskey CJ. Triggering germination represents a novel strategy to enhance killing of *Clostridium difficile* spores. PLoS One. 2010 Aug 19;**5**(8):e12285

[248] Piktel E, Pogoda K, Roman M, Niemirowicz K, Tokajuk G, Wroblewska M, et al. Sporicidal activity of ceragenin CSA-13 against *Bacillus subtilis*. Scientific Report. 2017 Mar;**7**:44452

[249] Finegold SM, Molitoris D, Vaisanen M, Song Y, Liu C, Bolaños M, et al. In vitro activities of OPT-80 and comparator drugs against intestinal bacteria. Antimicrobial Agents and Chemotherapy. 2004;**48**(12):4898-4902

[250] Babakhani F, Bouillaut L, Sears P, Sims C, Gomez A, Sonenshein AL. Fidaxomicin inhibits toxin production in *Clostridium difficile*. Journal of Antimicrobial Chemotherapy. 2013 Mar 1;**68**(3):515-522

[251] Babakhani F, Bouillaut L, Gomez A, Sears P, Nguyen L, Sonenshein AL. Fidaxomicin inhibits spore production in *Clostridium difficile*. Clinical Infectious Diseases. 2012 Aug 1;**55**(Suppl 2):S162–S169

[252] Allen CA, Babakhani F, Sears P, Nguyen L, Sorg JA. Both fidaxomicin and vancomycin inhibit outgrowth of *Clostridium difficile* spores. Antimicrobial Agents and Chemotherapy. 2013;**57**(1):664-667

[253] Mascio CTM, Mortin LI, Howland KT, Van Praagh ADG, Zhang S, Arya A, et al. In vitro and in vivo characterization of CB-183,315, a novel lipopeptide antibiotic for treatment of *Clostridium difficile*. Antimicrobial Agents and Chemotherapy. 2012;**56**(10):5023-5030

[254] Citron DM, Tyrrell KL, Merriam CV, Goldstein EJC. Comparative in vitro activities of LFF571 against *Clostridium difficile* and 630 other intestinal strains of aerobic and anaerobic bacteria. Antimicrobial Agents and Chemotherapy. 2012 May 6;**56**(5):2493-2503

[255] Snydman DR, Jacobus N V, McDermott LA. Activity of a novel cyclic lipopeptide, CB-183,315, against resistant *Clostridium difficile* and other gram-positive aerobic and anaerobic intestinal pathogens. Antimicrobial Agents and Chemotherapy. 2012 Jun 29;**56**(6):3448-3452

[256] Locher HH, Seiler P, Chen X, Schroeder S, Pfaff P, Enderlin M, et al. In vitro and in vivo antibacterial evaluation of cadazolid, a new antibiotic for treatment of clostridium difficile infections. Antimicrobial Agents and Chemotherapy. 2014;**58**(2):892-900

[257] Locher HH, Caspers P, Bruyère T, Schroeder S, Pfaff P, Knezevic A, et al. Investigations of the mode of action and resistance development of cadazolid, a new antibiotic for treatment of clostridium difficile infections. Antimicrobial Agents and Chemotherapy. 2014;**58**(2):901-908

[258] Goldstein EJC, Citron DM, Tyrrell KL, Merriam CV. Comparative in vitro activities of SMT19969, a new antimicrobial agent, against *Clostridium difficile* and 350 gram-positive and gram-negative aerobic and anaerobic intestinal flora isolates. Antimicrobial Agents and Chemotherapy. 2013;**57**(10):4872-4876

[259] Vickers RJ, Tillotson G, Goldstein EJC, Citron DM, Garey KW, Wilcox MH. Ridinilazole: A novel therapy for *Clostridium difficile* infection. International Journal of Antimicrobial Agents. 2016;**48**(2):137-143

[260] Bassères E, Endres BT, Khaleduzzaman M, Miraftabi F, Alam MJ, Vickers RJ, et al. Impact on toxin production and cell morphology in *Clostridium difficile* by ridinilazole

(SMT19969), a novel treatment for *C. difficile* infection. Journal of Antimicrobial Chemotherapy. 2016 May 1;**71**(5):1245-1251

[261] Critchley IA, Green LS, Young CL, Bullard JM, Evans RJ, Price M, et al. Spectrum of activity and mode of action of REP3123, a new antibiotic to treat *Clostridium difficile* infections. Journal of Antimicrobial Chemotherapy. 2009 May 1;**63**(5):954-963

[262] Citron DM, Warren YA, Tyrrell KL, Merriam V, Goldstein EJC. Comparative in vitro activity of REP3123 against *Clostridium difficile* and other anaerobic intestinal bacteria. Journal of Antimicrobial Chemotherapy. 2009 May 1;**63**(5):972-976

[263] Ochsner UA, Bell SJ, O'Leary AL, Hoang T, Stone KC, Young CL, et al. Inhibitory effect of REP3123 on toxin and spore formation in *Clostridium difficile*, and in vivo efficacy in a hamster gastrointestinal infection model. Journal of Antimicrobial Chemotherapy. 2009 May 1;**63**(5):964-971

[264] Atarashi K, Tanoue T, Oshima K, Suda W, Nagano Y, Nishikawa H, et al. Treg induction by a rationally selected mixture of Clostridia strains from the human microbiota. Nature. 2013 Aug 8;**500**(7461):232-236

[265] Narushima S, Sugiura Y, Oshima K, Atarashi K, Hattori M, Suematsu M, et al. Characterization of the 17 strains of regulatory T cell-inducing human-derived Clostridia. Gut Microbes. 2014;**5**(3):333-339

[266] Weingarden AR, Dosa PI, DeWinter E, Steer CJ, Shaughnessy MK, Johnson JR, et al. Changes in colonic bile acid composition following fecal microbiota transplantation are sufficient to control *Clostridium difficile* germination and growth. PLoS One. 2016;**11**(1):1-16

[267] Staley C, Weingarden AR, Khoruts A, Sadowsky MJ. Interaction of gut microbiota with bile acid metabolism and its influence on disease states. Applied Microbiology & Biotechnology. 2017;**101**(1):47-64

[268] Ott SJ, Waetzig GH, Rehman A, Moltzau-Anderson J, Bharti R, Grasis JA, et al. Efficacy of sterile fecal filtrate transfer for treating patients with *Clostridium difficile* infection. Gastroenterology. 2017 Mar;**152**(4):799-811.e7

[269] Villano SA, Seiberling M, Tatarowicz W, Monnot-Chase E, Gerding DN. Evaluation of an oral suspension of VP20621, spores of nontoxigenic *Clostridium difficile* strain M3, in healthy subjects. Antimicrobial Agents and Chemotherapy. 2012 Oct;**56**(10):5224-5229

[270] Nagaro KJ, Phillips ST, Cheknis AK, Sambol SP, Zukowski WE, Johnson S, et al. Nontoxigenic *Clostridium difficile* protects hamsters against challenge with historic and epidemic strains of toxigenic BI/NAP1/027 *C. difficile*. Antimicrobial Agents and Chemotherapy. 2013;**57**(11):5266-5270

[271] Gerding D, Meyer T, Lee C, Al E. Administration of spores of nontoxigenic *Clostridium difficile* strain m3 for prevention of recurrent *C. difficile* infection: A randomized clinical trial. The Journal of American Medical Association. 2015 May 5;**313**(17):1719-1727

[272] Brouwer MSM, Roberts AP, Hussain H, Williams RJ, Allan E, Mullany P. Horizontal gene transfer converts non-toxigenic *Clostridium difficile* strains into toxin producers. Nature Communication. 2013 Oct 17;**4**:2601

[273] Sougioultzis S, Kyne L, Drudy D, Keates S, Maroo S, Pothoulakis C, et al. *Clostridium difficile* toxoid vaccine in recurrent *C. difficile*-associated diarrhea. Gastroenterology. 2005;**128**(3):764-770

[274] Foglia G, Shah S, Luxemburger C, Pietrobon PJF. *Clostridium difficile*: Development of a novel candidate vaccine. Vaccine. 2012 Jun 19;**30**(29):4307-4309

[275] Donald RGK, Flint M, Kalyan N, Johnson E, Witko SE, Kotash C, et al. A novel approach to generate a recombinant toxoid vaccine against *Clostridium difficile*. Microbiology (United Kingdom). 2013;**159**(PART7):1254-1266

[276] Tian J-H, Fuhrmann SR, Kluepfel-Stahl S, Carman RJ, Ellingsworth L, Flyer DC. A novel fusion protein containing the receptor binding domains of *C. difficile* toxin A and toxin B elicits protective immunity against lethal toxin and spore challenge in preclinical efficacy models. Vaccine. 2012 Jun 13;**30**(28):4249-4258

[277] Permpoonpattana P, Hong HA, Phetcharaburanin J, Huang JM, Cook J, Fairweather NF, et al. Immunization with *Bacillus* spores expressing toxin a peptide repeats protects against infection with *Clostridium difficile* strains producing toxins A and B. Infection and Immunity. 2011;**79**(6):2295-2302

[278] Zackular JP, Moore JL, Jordan AT, Juttukonda LJ, Noto MJ, Nicholson MR, et al. Dietary zinc alters the microbiota and decreases resistance to *Clostridium difficile* infection. Nature Medicine. 2016 Nov;**22**(11):1330-1334

Chapter 3

Clostridium difficile Infection: Pathogenesis, Diagnosis and Treatment

Laura Fernández-García, Lucia Blasco,
María López and Maria Tomás

Additional information is available at the end of the chapter

http://dx.doi.org/10.5772/67754

Abstract

Clostridium difficile is a Gram-positive bacterium with the capacity of spore generation. The *C. difficile* infections, related to antibiotic treatment, have increased in number and severity during the last few years; increasing the health problems caused by this bacterium. One of the most important problems of the *C. difficile* infection is the recurrence. Due to all of these facts, researchers have been searching for new treatments such as faecal microbiota transplantation or bacteriocins development.

Keywords: clostridium, clinical, pathogenesis, persistence, resistance, treatment

1. Introduction

Clostridium difficile is a Gram-positive, spore-forming anaerobic bacterium discovered in 1935 by Hall and O'Toole [1]. In 1978, Barlett et al. identified *C. difficile* as an important cause of pseudomembranous colitis (PMC) associated with antibiotic use [2]. The manifestations of *C. difficile* infection (CDI) range from asymptomatic carriage to fulminant disease. Nonetheless, the commonest manifestations are diarrhoea and PMC [3]. One of the most serious problems associated with CDI is recurrence of the disease. *Clostridium difficile* infection can be acquired by person to person transmission, especially by the faecal-oral route, and it can also be acquired by environmental contamination [4]. *Clostridium difficile* is widely distributed in the soil and in the intestinal tracts of animals, both of which are considered as reservoirs of the bacterium [3].

Historically, CDI was not considered a severe disease. However, the number of cases and the severity of these have increased in the last 20 years [5]. One of the reasons for the increase in the incidence of CDI is that *C. difficile* produces spores that are capable of resisting heat, desiccation, and chemical agents. The appearance of *C. difficile* in the hospital environment has become problematical.

The incidence of CDI in hospitals depends on the type of unit, and the rates are highest in haematology, gastroenterology and nephrology units. The incidence also depends on the country considered, and within Europe, the rates are highest in Finland and Poland and lowest in Turkey, Bulgaria and in East European countries [6].

2. Clinical features

The most common symptom of CDI is watery (not bloody) diarrhoea accompanied by abdominal pain. Doctors should suspect CDI when the patient has three unformed or watery stools daily for 1 or 2 days. In case of more severe symptoms, the patient may present with fever, shock or hypotension and severe ileus with cessation of diarrhoea. The most severe symptoms are leucocytosis and elevated serum creatinine levels [7]. CDI may also lead to complications such as dehydration, electrolyte disturbance, hypoalbuminemia, toxic megacolon, bowel perforation, hypotension, renal failure, systemic inflammatory response syndrome, sepsis and death [4].

Colitis can affect any part of the colon but is commonly severest in the distal colon and rectum. Patients with CDI in these locations always present fever, abdominal pain, leucocytosis and a decrease in intestinal motility [3].

The most important risk factor for CDI is age, although the duration of hospitalization is also important, along with exposure to anti-microbial agents. Olson et al. showed that 96% of CDI cases had been exposed to anti-microbials about 14 days before the manifestation of diarrhoea, and all patients had received anti-microbial treatment about 3 months before [8]. Other investigators extend the time of influence of antibiotic treatment to 12 months and also include administration of proton pump inhibitors as a risk factor. Interestingly, the presence of diabetes mellitus has been associated with a decreased risk of CDI [6].

3. Pathogenic factors

3.1. Toxins

Clostridium difficile can produce three toxins: A, B and binary toxin. Toxins A and B were the first identified in this bacterium; both are encoded by genes in the pathogenicity locus (PaLoc) and are included in the large clostridial toxins (LCT), a family known to modify small GTPases [9]. These toxins act as glycosyltransferases that modify Rho and Ras proteins within the intestinal epithelial cells and disrupt the actin cytoskeleton, causing loss of intercellular junctions and the severe secretory diarrhoea associated with CDI [10, 11]. Hundsberger and collaborators reported that two of the LCT genes, *tcdD* and *tcdC*, act as positive and negative regulators, respectively. These researchers have classified the *tcd* genes in two groups, one comprising A, B, D and E, and the other comprising C (as *tcdC* has the opposite orientation). It has also been found that *tcdC* is expressed in the exponential growth phase of the bacterium, while the other genes are expressed in the stationary phase [12, 13].

Some 10% of *C. difficile* strains are capable of producing binary toxin. This toxin is classified as an ADP-ribosyltransferase and is encoded by the *cdtA* gene (the enzymatic component) and the *cdtB* gene (the binding component) [3, 9]. Binary toxin acts on the actin cytoskeleton, producing microtubule-based protrusions on the surface of epithelial cells [5]. A number of studies have indicated that strains that produce binary toxins usually cause severe CDI [14]. Geric et al. used a rabbit ileal loop model to investigate the binary toxin and concluded that the toxin contributes significantly to eliciting a non-haemorrhagic fluid response [15]. A higher mortality rate was observed in patients infected with strains that produce all three toxins [5].

Regulation of toxins A and B has been widely studied for many years, and it is known that the *tcdR* regulator gene, present in the PaLoc, activates *tcdA* and *tcdB* transcriptionally and also activates its own two promoters [16]. The four upstream genes of the PaLoc (*tcdA, B, E, R*) can be co-transcribed by the *tcdR* promoter, and each of the toxin genes has its own promoters. A global regulator of gene expression commonly found in low G+C Gram-positive bacteria, known as *CodY*, has been identified in *C. difficile*. Inactivation of *CodY* in a *C. difficile* strain has led to expression of the PaLoc genes during both exponential growth of the bacterium and in the stationary phase, demonstrating that *CodY* regulates toxin production. *CodY* has been shown to bind to the *tcdR* promoter with high affinity, especially when GTP and branched-chain amino acids are present; *CodY* also binds to toxin gene promoters, but with low affinity, which suggests that the primary regulation affects *tcdR* (**Figure 1**). *CodY* works by repressing

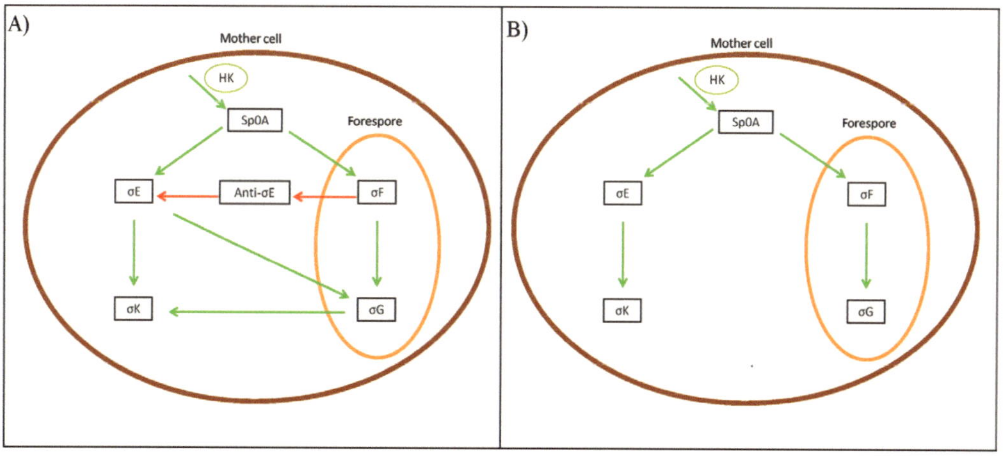

Figure 1. Regulation of sigma factors: (a) Regulation of the sigma factors in *Bacillus subtilis*, which has a criss-cross regulation between mother cell (where express factor E and K) and the forespore (where factors F and G have their function); factor F inhibits anti-σE factor which inhibits factor E. (b) Regulation of the sigma factors in *Clostridium difficile*, with its forespore inside the mother cell (like *Bacillus subtilis*).

some genes when nutrients are sufficient and de-repressing those genes when nutrients are limited. Researchers have discovered that *CodY* not only represses toxin genes but also represses *tcdC*; it is not known why *CodY* represses both toxins and its antagonist. It has been suggested that in the intestinal tract, where nutrients are not abundant, *CodY* may relax its repression of toxin genes so that the toxins will lyse epithelial cells in the intestinal tract, thus releasing nutrients [17].

Toxin production is affected both by the bacterial growth phase and by environmental factors. Researchers observed that the presence of glucose in the medium inhibits toxins, which implies catabolic repression. In addition, toxin production in laboratory cultures has also been found to be affected by the presence of biotin and some amino acids (cysteine and proline) and by environmental stress [16].

3.2. Persistence of spores

Spore production in bacteria is a mechanism of persistence, as it confers resistance to antibiotics and to the host immune system. Akerlund and colleagues observed an inverse relationship between toxin production and spore counts, which suggests that premature sporulation in the stationary phase shortens the time required for toxin production [18]. Merrigan et al. observed that some hypervirulent strains underwent early sporulation and produced large amounts of toxin, with greater efficiency than other strains. These researchers concluded that sporulation could contribute to the dissemination of infectious particles in the environment, thus helping toxins to confer adaptive advantages in the pathogenesis of hypervirulent strains of *C. difficile* [13].

The sporulation process has been widely studied in *Bacillus subtilis*. The sporulation decision in the genus Bacillus is regulated by some orphan histidine kinases whose function is to phosphorylate the master transcriptional regulator Spo0A [19]. Underwood et al. discovered that *C. difficile* also has five orphan histidine kinases and that inactivation of these significantly reduces spore formation related to wild-type. However, the Spo0A phosphorylation mechanism remains unclear [20].

Once a strain of *B. subtilis* has committed to sporulation (i.e. Spo0A has been phosphorylated), a cascade of activation of RNA polymerase sporulation-specific sigma factor occurs [21]. Recent genomic studies of *C. difficile* have shown that this bacterium does not have the characteristic criss-cross regulation of *B. subtilis* [22]. It is known that *B. subtilis* has four sigma factors (E, F, G and K) and that the active factor F has the capacity of inactivate anti-σ^E factor, thus enabling activation of factor E. Once factor E is activated, it can activate factor G, while at the same time factor G activates factor K. In addition, factor E is necessary for activation of factor K, while factor F is required for the activation of factor G. However, unlike *B. subtilis, C. difficile* uses factor F to activate post-translational factor G, and factor E activate factor F; but factor E is not necessary for activation of factor G, and factor G is also not necessary for activation factor K, as proteolytic activation is not useful in factor K [22]. The mechanism of regulation of all these factors is not known and is currently under study. However, Pereira et al. observed that as in *B. subtilis*, the activity of factors F and G in *C. difficile* is focused on forespores and that of factors E and K on the mother cell (**Figure 2**) [23].

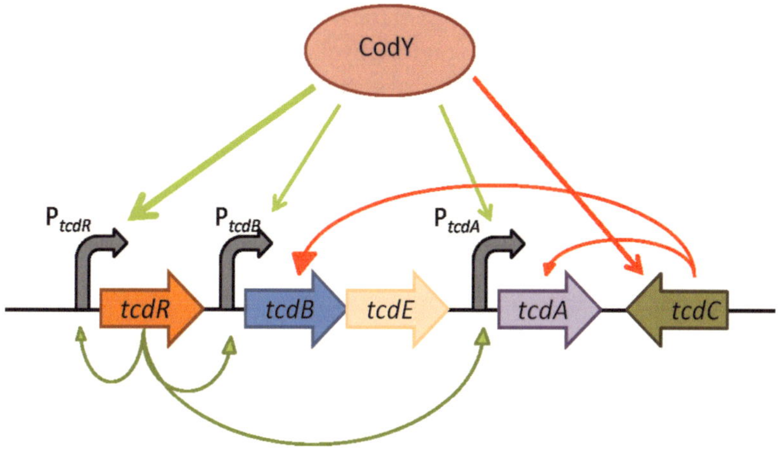

Figure 2. Regulation of toxins A and B in the PaLoc cluster. *tcdR* actives the promoters of *tcdB* and *tcdA*, meanwhile *tcdC* inhibits *tcdB* and *tcdA* genes; *tcdC* is also inhibit by CodY, till it active the tree promoters of the cluster.

Studies of the spore surface have identified surface receptors that interact with intestinal epithelial cells. *In vitro* experiments have shown that spores can become cytotoxic to macrophages and that spores disrupt the phagosomal membrane with the aid of their surface receptors [24].

Bacteria usually germinate when specific germinant receptors detect specific small molecules in the environment. However, *C. difficile* germinates after detecting some bile salts and L-glycine in the environment [25]. It is thought that CspC, a serine protease involved in germination in Clostridium perfringens, is a bile salt specific germinant receptor. It is known that CspC is required for Ca-DPA activation in spore germination in response to glycine and taurocholate [26].

Growth of *C. difficile* vegetative cells has been shown to be inhibited by cleaning agents and germicides, although the bacterium is not killed and undergoes sporulation. Spore formation makes it difficult to eliminate *C. difficile* and is one of the most common problems in the hospital environment. Several researchers are therefore investigating the use of cleaning agents and germicides to eliminate *C. difficile*. Fawley et al. found that neutral and hydrogen peroxide detergents do not reduce germination of spores, whereas chlorine-containing agents do reduce the rates of germination [27]. Destruction of *C. difficile* spores is very important, especially in healthcare environments, because if not destroyed, spores tend to accumulate and thus represent a potential health risk. Vapourized hydrogen peroxide has been shown to have an important sporicidal effect [28].

3.3. Resistance

Antibiotic resistance is a huge problem nowadays, especially with the appearance of new *C. difficile* ribotypes. The most dangerous of these is RT027, which is associated with excessive use of fluoroquinolones (FQs). The findings of the numerous studies concerning resistance of *C. difficile* have shown that resistance to erythromycin, fluoroquinolones and ciprofloxacin is very common in clinical strains. *C. difficile* also generally displays resistance to second-generation cephalosporins but shows less resistance to third-generation cephalosporins. However, resistance to moxifloxacin and gatifloxacin has been detected in 34% of strains analysed [29]. Tenover and colleagues analysed the resistances of different ribotypes to clindamycin, metronidazole, moxifloxacin and rifampin. They observed that resistance rates are changed between strain types as well as in Europe and Far East [30].

Clostridium difficile has a number of putative β-lactamase genes that are probably involved in resistance to β-lactam antibiotics [29]. A ribosomal methylation mechanism related to macrolide resistance in *C. difficile* has been described (Ref). Erythromycin ribosomal methylases (ERM) confer resistance to macrolides, and some of these genes have been described in *C. difficile* [31]. These genes are widespread in clinical strains of *C. difficile*, despite the fitness cost associated with maintaining them.

Alterations in the quinolone-resistance determining region (QRDR), which confers resistance to fluoroquinolones, have been identified in *C. difficile* [32]. Resistance of *C. difficile* to tetracycline varies in different countries, ranging from 2.4 to 41.67% [32]. Fry et al. reported that although the tetracycline resistance gene *tetM* is predominant in *C. difficile*, others such as *tetW* are found in human and animal strains [33]. The presence of other mobile elements in the genome involved in resistance to tetracycline is also possible [33]. Freeman and colleagues investigated chloramphenicol resistance in clinical isolates of *C. difficile* and attributed it to the presence of *catD* gene, which encodes a chloramphenicol acetyltransferase that can be found in transposons [34].

3.4. Hypervirulent strains

Some hypervirulent strains of *C. difficile* have appeared in recent years, representing a huge health problem. All strains of *C. difficile* can be classified into 150 ribotypes and 24 toxinotypes. The toxinotypes are classified on the basis of different poly-morphisms in the PaLoc. All members of toxinotype III, to which ribotypes 027, 034, 075, 080 belong, produce binary toxin [3].

Merrigan and collaborators have shown that early sporulation in hypervirulent strains enables accumulation of more spores than in non-hypervirulent strains and thus explains the incidence of recurrent infection associated with hypervirulent strains [13]. Furthermore, the high rate of CDI has been related to the higher rate of toxin production in these strains than in non-hypervirulent strains. Yakob et al. made an epidemiological model of *C. difficile* transmission, in this model they could observe that hypervirulent strains seem to be more infectious, more likely to become established, extend faster and have a higher presence in the community, thus displacing endemic strains. These statements are based on the fact that in the last 15 years hypervirulent strains (especially ribotype 027) not only have appeared but also have become in the dominant strain worldwide [35].

http://dx.doi.org/10.5772/67754

Ribotype 027 is one of the most dangerous and best studied strains of *C. difficile*. This strain produces more A and B toxins than 'normal' strains and also produces binary toxin. Although ribotype 027 was first identified in Canada, it has extended throughout the world and is now endemic in the United States [7, 36]. The virulence of *C.difficile* ribotype 027 has been suggested to be due to delection in position 117 of the *tcdC* gene which produces an increase in toxin production and enables binary toxin production [37]. This strain is commonly associated with severe cases of CDI and high rates of recurrence, with elevated mortality.

Another hypervirulent strain has recently become problematical in Europe, especially in healthcare environments. Ribotype 078, which has been identified in many cases of CDI throughout Europe, shows similar hyperproduction of toxins as in ribotype 027 [37].

3.5. Diagnosis

CDI can be diagnosed by detection of genes or products or by bacterial culture. Culture of *C. difficile* takes at least 4 days to be detected on *C. difficile* plates. Culture strains are useful for typifying *C. difficile* and allow to store it for future research.

Different tests can be used to detect products. The cell cytotoxicity assay (CCA) is the gold standard assay for detecting CDI. There are also other more sensitive methods, such as the toxigenic culture method. This method involves culturing *C. difficile* in selective media and subsequent demonstration of toxin production by ELISA or CCA. However, the most commonly used test is detection of *C. difficile* toxins by enzyme immunoassay (EIA), either directly or with glutamate dehydrogenase antigen. This assay displays sensitivity of 63–94% and specificity of 75–100%. There is another EIA test against a common *C. difficile* antigen, glutamate dehydrogenase (GDH), which can also be used. This test displays 58–68% sensitivity and 94–98% specificity. Although this method is not sufficiently sensitive for routine laboratory use, it is useful for epidemiological research [4].

Real-time PCR is currently the fastest available test for CDI and is usually used to detect genes regulating synthesis of toxins A and B. The tests used must be capable of distinguishing between colonization and disease [3].

CDI can also be diagnosed by direct visualization, especially in cases of PMC. Nonetheless, direct visualization only detects CDI in 51–55% of cases and laboratory tests must be conducted to confirm the diagnosis [4].

4. Classical treatments

Treatment of CDI depends on whether the disease is classified as first episode, recurrent, or severe or complicated CDI. The treatment of choice for first episode CDI is usually metronidazole. However, other agents such as rifaximin and teicoplanin can be used. Rifaximin, a nonabsorbable oral antibiotic, is effective against first and recurrent episodes of CDI. Teicoplanin, which is similar to vancomycin, is not approved for use in the US [10].

Administration of vancomycin is recommended in cases of recurrent CDI. Metronidazole is not used in the recurrent episodes or as long-term therapy because of its potential neurotoxicity. Fidaxomicin, a novel macrocyclic antibiotic, has appeared in recent years and doctors are being considered it as a substitute for vancomycin [4, 10].

For patients with severe or complicated CDI, vancomycin can be administered into the colon, or metronidazole can be administered intravenously. However, oral or rectal administration of vancomycin is recommended. In the most serious cases, or when there is no antibody response to antibodies, colectomy may be the best solution to save patients' lives. The surgery is only performed on patients with megacolon, colonic perforation, acute abdomen or septic-shock [4]. An alternative to colectomy is diverting loop ileostomy followed by an intraoperative lavage with 8 L of poly-ethylene glycol and 500 mg vancomycin every 8 hours [7].

5. New treatments

5.1. Monoclonal antibodies

Some researchers have tested the efficacy of intravenous monoclonal antibodies in preventing recurrence of CDI. Patients who were first treated with metronidazole or vancomycin received treatment with antibodies against toxin A (CDA1) and toxin B (CDB1) in a double-blind experiment. The results of the study showed that relative to control patients, those receiving the antibody treatment showed a lower rate of recurrence of CDI in the 12 weeks of the study [7].

5.2. Faecal microbiota transplantation (FMT)

This method consists of transferring a suspension of faecal matter from a healthy donor to a patient with the aim of recovering the 'normal' microbiota. The donor material is introduced into the patient via rectal enema, a nasoduodenal tube or colonoscopy. An FMT is recommended after a third CDI episode [38]. Kelly et al. conducted a randomized trial, in which they selected adults with three or more documented CDI recurrences. The faecal microbiota from patients and donors was analysed 5 days before and some weeks after the treatment. The first analysis revealed that the patients had more gammaproteobacteria and betaproteobacteria and fewer firmicutes and Bacteroidetes than donors. They concluded that administration of fresh FMT from a donor via colonoscopy to patients who were first administered a course of vancomycin was successful in preventing further CDI episodes. The researchers also observed that the efficacy of the treatment varied depending on the part of the intestine where the infection occurred, and therefore some patients may not benefit from FMT [39].

Although this treatment has been shown to be successful in some cases, some adverse events also occurred due to the impracticability of screening all possible pathogens carried by the donor. The long-term consequences of the treatment are also unpredictable. The US Food and Drug Administration (FDA) has classified FMT as a drug and biological product, so that the procedure is subjected to the same regulations as traditional pharmaceutical drugs [38].

5.3. Spore formulation

This treatment is similar to FMT, although in this case, a specific number of strains compete with *C. difficile* [7]. Khanna et al. produced a mixture of spores of 50 species of Firmicutes obtained from healthy donors. After some success in preclinical studies, the mixture, known as ser-109, was formulated for oral administration to patients between 18 and 90 years with recurrent CDI. The researchers observed a reduction at recurrence of the CDI of 87.7% over 8 weeks of treatment [40]. Gerding et al. recently began a phase 2 trial of administration of spores from a single non-toxigenic *C. difficile*, in patients who had responded to antibiotics in the first episode or the first recurrence [41]. Positive results of the studies support the use of spores as a treatment against recurrent CDI, although further detailed studies are required.

5.4. Whole genome sequencing (WGS)

Whole genome sequencing (WGS) provides high-resolution data, enabling researchers to identify the strains of *C. difficile* isolated from patients and thus to distinguish between reinfection and relapse and to help to understand the complex transmission epidemiology of CDI. WGS can also be used to construct a transmission map. Researchers have used WGS and other similar techniques to elucidate CDI transmission events based on culture of isolates from patients with CDI [42].

WGS has been used to map the transmission of strains through patients and across countries [43]. For example, various genetically closely related strains have been identified in Australia. The researchers hypothesized that an animal vector is the cause of this expansion. An Australian strain has been identified in England, in a patient who had was previously visited Australia [44].

A new sequencing method known as Oxford Nanopore's MinION sequencer (http://www.nanoporetech.com/) has been shown to be potentially useful for CDI fast diagnosis [7].

5.5. Microbiome-wide association studies (MWAS)

MWAS are carried out to help in understanding the interactions between bacteria within their communities and to discover 'which produce infections and why' [45]. Koenigsknecht et al. showed that the microbiome modifies bile-acid metabolite profiles during establishment of *C. difficile* in mice [46]. The study involved the use of metagenomics, metatranscriptomics, metaproteomics and metabolomics to determine how microbiota helps the host fight against CDI. Clostridium scindens was identified by MWAS as a candidate for fighting against *C. difficile* [45]. Allegretti et al. found that some bile acids are involved in resistance to *C. difficile* during treatment with antibiotics [47]. MWAS is proved useful in predicting responses to treatment or the development of disease [45].

5.6. Probiotics

Probiotics are defined as live microorganisms, which, when administered in adequate amounts, confer a health benefit to the host [48]. Probiotics have three modes of action: (i) modulation of host defences; (ii) effects on other microorganisms. Probiotic bacteria adhere to epithelial cells, which block adherence of pathogens. Thus, if pathogens cannot adhere to epithelium they cannot invade the cells and (iii) effects on microbial products such as toxins or host products. Some microorganisms can inhibit toxin production by producing other toxins. For example, the presence of *Saccharomyces boulardii* provides some protection against *C. difficile* toxin A [49].

Tung et al. conducted a review of the use of S. boulardii to treat diarrhoea. These researchers showed that *S. boulardii* plays an important role in preventing both primary and recurrent CDI [50]. They conducted an observational study of the efficacy of a probiotic mixture (containing Lactobacillus acidophilus, Lactobacillus casei and Lactobacillus rhamnosus); the probiotic was administered about 2–12 hours after antibiotic for 30 days or until the end of the treatment. A reduction of 39% of CDI cases was observed over the following 10 years [51].

Bioengineering of microorganisms to target specific pathogens has gained popularity in recent years [38]. Up to date, no research has been carried out in this field in relation to *C. difficile*.

5.7. Small molecule inhibitors

Many small molecules in the human body are capable of interfering with cellular processes by inhibiting or enhancing them. Researchers have found that both types of *C. difficile* toxins have a putative binding domain, which is a cysteine protease domain (CPD) and a glucosyltransferase domain (GTD). When bacteria find IP6 (1D-myo-inositol hexakisphosphate), the CPD activates GTD and produces toxicity [52].

In a study carried out with the aim of finding an inhibitor of TcdB CPD activator, Bender and colleagues discovered 44 inhibitors, the most promising of which was ebselen (phase 2 clinical trials). Ebselen is a synthetic low weight compound that is capable of reducing oxidative stress. Ebselen has been shown to inhibit CPD by blocking binding to IP6, which implies inhibition of toxic effects of TcdB *in vivo* [53]. Both *TcdA* and *TcdB* are regulated by thiolactone molecule, so that if the inhibitor of thiolactone can be identified, it should be possible to create a non-antibiotic treatment for CDI [54].

5.8. Bacteriocins

Bacteriocins are anti-microbial peptides produced by bacteria. Although a number of bacteriocins against *C. difficile* have been identified, three in particular appear to be the most effective: lacticin 3147, nisin and thuricin D.

Lacticin 3147 is a two component antibiotic produced by Lactobacillus lactis [55]. This compound is active at physiological pH, unlike nisin. Rea et al. demonstrated that lacticin can clear a broth of *C. difficile* when added during exponential growth of the bacterium. They also demonstrated that lacticin does not affect non-spore-forming Gram-negative bacteria, but it can reduce the presence of Gram-positive bacteria such as enterococci, lactobacilli and bifidobacteria. These researchers also showed that lacticin can kill *C. difficile* in a model faecal environment. As lacticin cannot resist gastric transit, administration via enema is recommended [56].

Thuricin CD is a two component agent belonging to the sactibiotic subclass of bacteriocins. It is produced by Bacillus thuringiensis and has been demonstrated to be more effective than vancomycin against *C. difficile*. Thuricin CD displays a potent activity against *C. difficile* and its activity against ribotype 027 has been highlighted. The capacity of thuricin CD to effectively kill *C. difficile* in a model of the distal human colon has been demonstrated [57]. However, like other bacteriocins, thuricin CD may not survive gastric transit, especially because one of its compounds is particularly susceptible to degradation. However, Rea et al. have demonstrated that thuricin CD administrated via the rectal route was effective in reducing CDI symptoms [58].

Nisin is a polypeptide of 34 aminoacid residues produced by Lactococcus lactis subspecies and inhibits a wide range of pathogens. Unlike lacticin and thuricin CD, nisin is classified as GRAS (generally recognized as safe) and can therefore be used as a food additive. Lay et al. determined the MICs of nisin A and showed that this bacteriocin is at least as effective against *C. difficile* as vancomycin. These researchers also demonstrated that nisin can inhibit the growth of *C. difficile* after germination, but that is not able to inhibit spores [59].

Acknowledgements

Funding: Laura Fernández-García was financially supported by a predoctoral fellowship from the Xunta de Galicia (GAIN, Axencia de Innovación). M. Tomás was financially supported by the Miguel Servet Research Programme (SERGAS and ISCIII). This chapter was funded by grant PI13/02390 and PI16/01163 awarded to M. Tomás within the State Plan for R+D+I 2013-2016 (National Plan for Scientific Research, Technological Development and Innovation 2008-2011) and co-financed by the ISCIII-Deputy General Directorate of evaluation and Promotion of Research - European Regional Development Fund "A way of Making Europe" and Instituto de Salud Carlos III FEDER, Spanish Network for the Research in Infectious Diseases (REIPI RD12/0015).

Author details

Laura Fernández-García, Lucia Blasco, María López and Maria Tomás*

*Address all correspondence to: ma.del.mar.tomas.carmona@sergas.es

Department of Microbiology, A Coruña Hospital (CHUAC)-INIBIC, A Coruña, Spain

References

[1] Hall I.C., O'Toole E. Intestinal flora in new-born infants: With a description of a new pathogenic anaerobe, Bacillus difficilis. Am J Dis Child. 1935;**49**(2):390-402. DOI: 10.1001/archpedi.1935.01970020105010

[2] Bartlett J.G., Chang T.W., Gurwith M., Gorbach S.L., Onderdonk A.B. Antibiotic-associated pseudomembranous colitis due to toxin-producing clostridia. N Engl J Med. 1978;**298**(10):531-534. DOI: 10.1056/NEJM197803092981003

[3] Kuijper E.J., Coignard B., Tull P. Emergence of *Clostridium difficile*-associated disease in North America and Europe. Clin Microbiol Infect. 2006;**12**(6):2-18. DOI: 10.1111/j.1469-0691.2006.01580.x

[4] Cohen S.H., Gerding D.N., Johnson S., Kelly C.P., Loo V.G., McDonald L.C., et al. Clinical practice guidelines for *Clostridium difficile* infection in adults: 2010 update by the society for healthcare epidemiology of America (SHEA) and the infectious diseases society of America (IDSA). Infect Control Hosp Epidemiol. 2010;**31**(5):431-455. DOI: 10.1086/651706

[5] Bacci S., Molbak K., Kjeldsen M.K., Olsen K.E. Binary toxin and death after *Clostridium difficile* infection. Emerg Infect Dis. 2011;**17**(6):976-982. DOI: 10.3201/eid/1706.101483

[6] Kurti Z., Lovasz B.D., Mandel M.D., Csima Z., Golovics P.A., Csako B.D., et al. Burden of *Clostridium difficile* infection between 2010 and 2013: Trends and outcomes from an academic center in Eastern Europe. World J Gastroenterol. 2015;**21**(21):6728-6735. DOI: 10.3748/wjg.v21.i21.6728

[7] Gerding D.N., File T.M., McDonald L.C. Diagnosis and treatment of *Clostridium difficile* Infection. Infect Dis Clin Prac. 2016;**24**(2):3-10. DOI: 10.1097/ipc.0000000000000382

[8] Olson M.M., Shanholtzer C.J., Lee J.T., Gerding D.N. 10 years of prospective clostridium-difficile-associated disease surveillance and treatment at the minneapolis-va-medical-center. 1982-1991. Infect Control Hosp Epidemiol. 1994;**15**(6):371-381.

[9] Davies A.H., Roberts A.K., Shone C.C., Acharya K.R. Super toxins from a super bug: Structure and function of *Clostridium difficile* toxins. Biochem J. 2011;**436**(3):517-526. DOI: 10.1042/bj20110106

[10] Ananthakrishnan A.N. *Clostridium difficile* infection: epidemiology, risk factors and management. Nat Rev Gastroenterol Hepatol. 2011;**8**(1):17-26. DOI: 10.1038/nrgastro.2010.190

[11] Kuehne S.A., Cartman S.T., Heap J.T., Kelly M.L., Cockayne A., Minton N.P. The role of toxin A and toxin B in *Clostridium difficile* infection. Nature. 2010;**467**(7316):711-713. DOI: http://www.nature.com/nature/journal/v467/n7316/abs/nature09397.html#supplementary-information

[12] Hundsberger T., Braun V., Weidmann M., Leukel P., Sauerborn M., von Eichel-Streiber C. Transcription analysis of the genes tcdA-E of the pathogenicity locus of *Clostridium difficile*. Eur J Biochem. 1997;**244**(3):735-742.

[13] Merrigan M., Venugopal A., Mallozzi M., Roxas B., Viswanathan V.K., Johnson S., et al. Human hypervirulent *Clostridium difficile* strains exhibit increased sporulation as well as robust toxin production. J Bacteriol. 2010;**192**(19):4904-4911. DOI: 10.1128/jb.00445-10

[14] Barbut F., Gariazzo B., Bonne L., Lalande V., Burghoffer B., Luiuz R., et al. Clinical features of *Clostridium difficile*-associated infections and molecular characterization

of strains: results of a retrospective study, 2000-2004. Infect Control Hosp Epidemiol. 2007;**28**(2):131-139. DOI: 10.1086/511794

[15] Geric B., Carman R.J., Rupnik M., Genheimer C.W., Sambol S.P., Lyerly D.M., et al. Binary toxin-producing, large clostridial toxin-negative *Clostridium difficile* strains are enterotoxic but do not cause disease in hamsters. J Infect Dis. 2006;**193**(8):1143-1150. DOI: 10.1086/501368

[16] Karlsson S., Burman L.G., Akerlund T. Induction of toxins in Clostridium difficile is associated with dramatic changes of its metabolism. Microbiology. 2008; 154 (Pt11):3430-3436. DOI: 10.1099/mic.0.2008/019778-0

[17] Dineen S.S., Villapakkam A.C., Nordman J.T., Sonenshein A.L. [dissertation]. 2007. DOI: 10.1111/j.1365-2958.2007.05906.x

[18] Akerlund T., Svenungsson B., Lagergren A., Burman L.G. [dissertation]. 2006. DOI: 10.1128/jcm.44.2.353-358.2006

[19] Higgins D., Dworkin J. [dissertation]. 2012. DOI: 10.1111/j.1574-6976.2011.00310.x

[20] Underwood S., Guan S., Vijayasubhash V., Baines S.D., Graham L., Lewis R.J., et al. [dissertation]. 2009. DOI: 10.1128/JB.00882-09

[21] de Hoon M.J., Eichenberger P., Vitkup D. [dissertation]. 2010. DOI: 10.1016/j.cub.2010.06.031

[22] Fimlaid K.A., Bond J.P., Schutz K.C., Putnam E.E., Leung J.M., Lawley T.D., et al. [dissertation]. 2013. DOI: 10.1371/journal.pgen.1003660

[23] Pereira F.C., Saujet L., Tomé A.R., Serrano M., Monot M., Couture-Tosi E., et al. [dissertation]. 2013. DOI: 10.1371/journal.pgen.1003782

[24] Paredes-Sabja D., Cofre-Araneda G., Brito-Silva C., Pizarro-Guajardo M., Sarker M.R. [dissertation]. 2012. DOI: 10.1371/journal.pone.0043635

[25] Paredes-Sabja D., Shen A., Sorg J.A. [dissertation]. 2014. DOI: 10.1016/j.tim.2014.04.003

[26] Francis M.B., Allen C.A., Shrestha R., Sorg J.A. [dissertation]. 2013. DOI: 10.1371/journal.ppat.1003356

[27] Fawley W.N., Underwood S., Freeman J., Baines S.D., Saxton K., Stephenson K., et al. [dissertation]. 2007. DOI: 10.1086/519201

[28] Otter J.A., French G.L., Adams N.M.T., Watling D., Parks M.J. [dissertation]. 2006. DOI: 10.1016/j.jhin.2005.08.011

[29] Spigaglia P. [dissertation]. 2016. DOI: 10.1177/2049936115622891

[30] Tenover F.C., Tickler I.A., Persing D.H. Antimicrobial-resistant strains of *Clostridium difficile* from North America. Antimicrob Agents Chemother. 2012;**56**(6):2929-2932. DOI: 10.1128/AAC.00220-12

[31] Spigaglia P., Carucci V., Barbanti F., Mastrantonio P. [dissertation]. 2005. DOI: 10.1128/aac.49.6.2550-2553.2005

[32] Dridi L., Tankovic J., Burghoffer B., Barbut F., Petit J.C. [dissertation]. 2002. DOI: 10.1128/aac.46.11.3418-3421.2002

[33] Fry P.R., Thakur S., Abley M., Gebreyes W.A. [dissertation]. 2012. DOI: 10.1128/jcm.06581-11

[34] Freeman J., Vernon J., Morris K., Nicholson S., Todhunter S., Longshaw C., et al. [dissertation]. 2015. DOI: 10.1016/j.cmi.2014.09.017

[35] Yakob L., Riley T.V., Paterson D.L., Marquess J., Magalhaes R.J.S., Furuya-Kanamori L., et al. [dissertation]. 2015. DOI: 10.1038/srep12666

[36] Pepin J., Valiquette L., Alary M.E., Villemure P., Pelletier A., Forget K., et al. [dissertation]. 2004. DOI: 10.1503/cmaj.1041104

[37] Kuijper E.J., Barbut F., Brazier J.S., Kleinkauf N., Eckmanns T., Lambert M.L., et al. [dissertation]. 2008.

[38] Culligan E.P., Sleator R.D. [dissertation]. 2016. DOI: 10.3390/jcm5090083

[39] Kelly C.R., Khoruts A., Staley C., Sadowsky M.J., Abd M., Alani M., et al. [dissertation]. 2016. DOI: 10.7326/M16-0271

[40] Khanna S., Pardi D.S., Kelly C.R., Kraft C.S., Dhere T., Henn M.R., et al. [dissertation]. 2016. DOI: 10.1093/infdis/jiv766

[41] Gerding D.N., Meyer T., Lee C., Cohen S.H., Murthy U.K., Poirier A., et al. [dissertation]. 2015. DOI: 10.1001/jama.2015.3725

[42] Gerding D.N. [dissertation]. 2016. DOI: 10.1093/cid/civ1037

[43] Mac Aogain M., Moloney G., Kilkenny S., Kelleher M., Kelleghan M., Boyle B., et al. [dissertation]. 2015. DOI: 10.1016/j.jhin.2015.01.021

[44] Eyre D.W., Tracey L., Elliott B., Slimings C., Huntington P.G., Stuart R.L., et al. [dissertation]. 2015.

[45] Gilbert J.A., Quinn R.A., Debelius J., Xu Z.Z., Morton J., Garg N., et al. [dissertation]. 2016. DOI: 10.1038/nature18850

[46] Koenigsknecht M.J., Theriot C.M., Bergin I.L., Schumacher C.A., Schloss P.D., Young V.B. [dissertation]. 2015. DOI: 10.1128/iai.02768-14

[47] Allegretti J.R., Kearney S., Li N., Bogart E., Bullock K., Gerber G.K., et al. [dissertation]. 2016. DOI: 10.1111/apt.13616

[48] FAO/WHO. [dissertation]. 2001.

[49] Oelschlaeger T.A.. [dissertation]. 2010. DOI: 10.1016/j.ijmm.2009.08.005

[50] Tung J.M., Dolovich L.R., Lee C.H. [dissertation]. 2009.

[51] Maziade P.J, Pereira P., Goldstein E.J. [dissertation]. 2015. DOI: 10.1093/cid/civ178

[52] Shen A., Lupardus P.J., Gersch M.M., Puri A.W., Albrow V.E., Garcia K.C., et al. [dissertation]. 2011. DOI: 10.1038/nsmb.1990

[53] Bender K.O., Garland M., Ferreyra J.A., Hryckowian A.J., Child M.A., Puri A.W., et al. [dissertation]. 2015. DOI: 10.1126/scitranslmed.aac9103

[54] Darkoh C., DuPont H.L., Norris S.J., Kaplan H.B. [dissertation]. 2015. DOI: 10.1128/mBio.02569-14

[55] McAuliffe O., Ryan M.P., Ross R.P., Hill C., Breeuwer P., Abee T. [dissertation]. 1998.

[56] Rea M.C., Clayton E., O'Connor P.M., Shanahan F., Kiely B., Ross R.P., et al. [dissertation]. 2007. DOI: 10.1099/jmm.0.47085-0

[57] Rea M.C., Dobson A., O'Sullivan O., Crispie F., Fouhy F., Cotter P.D., et al. [dissertation]. 2011. DOI: 10.1073/pnas.1001224107

[58] Rea M.C., Alemayehu D., Casey P.G., O'Connor P.M., Lawlor P.G., Walsh M., et al. [dissertation]. 2014. DOI: 10.1099/mic.0.068767-0

[59] Le Lay C., Dridi L., Bergeron M.G., Ouellette M., Fliss I. [dissertation]. 2016. DOI: 10.1099/jmm.0.000202

Chapter 4

Clostridium difficile Infection Diagnosis by Biological Molecular Methods

Luminița Smaranda Iancu, Andrei Florin Cârlan and
Ramona Gabriela Ursu

Additional information is available at the end of the chapter

http://dx.doi.org/10.5772/intechopen.68692

Abstract

In the past 15 years, the incidence of *Clostridium difficile* infection has emerged especially because of the new highly virulent strains. The classical diagnosis methods used to diagnose *C. difficile* infection take time and the enzyme immunoassay (EIA) test has demonstrated the lack of sensitivity. Even though new modern molecular methods have become available, the diagnosis of *C. difficile* in patients or healthy carriers remains a big challenge for both clinicians and laboratory staff. In the present chapter, we will list the main genotyping methods, stressing their advantages and disadvantages, as well. A brief presentation of the most useful kit (principle, sensitivity, specificity, benefits and disadvantages) to assess the impact of molecular methods in comparison with classical methods will offer support for future research in the present context of an increasing prevalence of *C. difficile* infection that represents worldwide, a real public health problem. To improve the patients' quality of life, to limit hospital transmission, and to save money, we have tried to identify the best diagnosis algorithm as tool in *C. difficile* diagnosis and surveillance. This algorithm may differ depending on the capacities of the laboratories and on the socioeconomic level of the countries in question.

Keywords: *C. difficile*, molecular method, PCR, RT-PCR, REA, PFGE, MLVA, VNRT, MLST, typing, diagnosis algorithm

1. Introduction

In the last years, especially starting with the early 2000s, the *Clostridium difficile* infection (CDI) has become a top subject in the medical field all over the world because of both high prevalence of healthcare-associated infection (HAI) and community infection. The causes of these spectra, including colonization and the diversity of clinical pictures, are topics of different chapters and, for this reason, we approach the subject directly. For a better understanding of the biological methods, the readers must look first at the genetics chapter.

The growing incidence of the infection with *C. difficile* around the globe, especially because of the highly virulent strains, has changed the approach of laboratory diagnosis and imposed the drafting of numerous guides, which include preventive measures to stop their circulation [1]. Different expert groups, published in the last years large studies on the national programs for *C. difficile* surveillance, coming to the conclusion that many countries do not have the capacity to quickly diagnose the *C. difficile* infections, and that hospital outbreaks led to a continuous spread of the strains, including the highly virulent ones [2]. In countries as Romania, Bulgaria, and Greece, in which the Eurosurveillance studies (run by EUCLID) concluded the low-level capacity of laboratories, in the last years, at national level, the experts have initiated the implementation of a new guide to improve the surveillance of *C. difficile* [2, 3]. The apparent low incidence of CDI in some countries, such as China, is the consequence of the limited laboratory diagnosis capacities [4].

The overuse of antibiotics and the long and repeated hospitalization, especially in the case of patients over 65, are the main causes of this increase [5]. This change is mainly related to the high diversity of virulent strains and to the new approach of the antibiotic therapeutic scheme [6]. A long list of antibiotics, including the large broad spectrum, may be the cause of the dysbiosis that lead to CDI: clindamycin, cephalosporins (e.g., cefaclor, cefotaxime), β-lactams (e.g., penicillins, ampicillin, amoxicillin-clavulanic acid), fluoroquinolones, including last generations of gatifloxacin, levofloxacin, and moxifloxacin [5, 7, 8]. Not all authors accept the whole list; some of them consider cephalosporins to be certainly involved, whereas others consider fluoroquinolones (cited by Daniel and Rapose [9]). CDI was noticed even after short-term antibiotic courses, inclusively as a preventive measure [5]. Since 2002, the most virulent strains of *C. difficile*, described in literature as toxinotype IIINAP1/027, were isolated as resistant to fluoroquinolones [7]. The 027 toxinotype nomination is related to laboratory methods, pulsed field gel electrophoresis (PFGE), and to the place of isolation (North American PFGE type 1 (NAP1), respectively). This resistance pattern is associated with the ability to produce A and B toxins in larger quantities, including binary toxin production, and has a greater capacity to spread endemically [6].

The Center for Disease Control and Prevention (USA, Atlanta) consider CDI as the first cause of healthcare-associated infection [10]. All authors agree that CDIs are related to *C. difficile* strains, which produce toxins (A, B, and binary toxin, and many other virulence factors) [11, 12].

With the new virulence strains that have emerged, the incidence, morbidity, and mortality through CDI have increased all over the world, especially in the developed countries, starting with the early 2000s [13]. In USA, CDC estimated that a quarter million people need hospitalization and around 5.6% die from CDI each year; more than 90% of deaths occur in people over 65 years, but almost 50% of the infections were noted in patients younger than 65 [10]. In parallel with the overuse of antibiotics, many old patients have co-morbidities that increase the risk of severe evolution; on the other hand, the need for quick, sensitive, and specific diagnosis have led to new laboratory techniques, including high-level biological and molecular ones that improved the positivity rate, decreasing the false negative one. All these factors, together with the previously mentioned and, probably, with other factors (e.g., pump protons inhibitors) made that, only from 2000 to 2007, the death number related to CDI to increase with 400% [10]. In the intensive care unit (ICU), the prevalence of CDI is the highest because it cumulates many risk factors. An excellent meta-analysis, from 2015, concluded that the prevalence of CDI among diarrheic ICU patients is more than five times higher in comparison with ICU patients with risk for pseudomembranous colitis, after approximately 10.5 days. The mortality rate was also higher: 32 versus 24%. This higher morbidity and mortality rate from ICU requires additional measures in order to prevent the spread of infections and more expensive regimens [14].

Even though the number of studies conducted to find out the costs of medical burden have increased, only a few of them have considered all factors that influence healthcare-associated costs; briefly, the CDC considered "at least $1 billion in excess medical costs per year" [10], and Kyne et al. estimated that *C. difficile* diarrhea cost/case was around $4600 and, using the appropriate statistics analysis, concluded that, in the USA, the total annual cost for diarrhea treatment exceeds $1.1 billion [15]. Some authors have considered different algorithms for diagnosis using phenotypic and biological molecular methods, concluding that the modern methods are less expensive than the traditional diagnostic ones recommended by some guides [16]. The risk of rapid spread in healthcare facilities of highly virulent strains, including those resistant to antibiotics, has led to improved laboratory techniques that have become able to quickly highlight CDI, with the possibility of applying the most appropriate preventive measures and modified therapeutic schemes [17, 18].

2. Arguments for the rapid diagnosis of *C. difficile* infection/colonization by molecular methods

The rapid diagnosis by molecular methods is costly at first sight, when compared to the rapid detection of toxins by enzyme immunoassay (EIA) or glutamate dehydrogenase (GDH) detection (an antigen produced in high amounts by all *C. difficile* strains, including the non-toxigenic one). Even though the EIA is a cheap test and despite the different commercial variants available, the specificity varied from 40 to 100% [13], with no major consequences regarding the clinical impact because a positive NAAT cannot differentiate between infection and colonization (**Table 1**). For this reason, it is necessary to also test the biomarkers that suggest the active infection. Different biomarkers were used, e.g., fecal lactoferrin and calprotectin, or cytokine analysis, but they did not demonstrate an efficient support in this differentiation [13].

In the modern era, when molecular platforms play an important role in patient diagnosis and management cases, a clear approach must be taken into consideration in order to use this tool in the best way, to reach as soon as possible a sensitive and specific diagnosis for patients and to reveal the most useful epidemiological markers, to initiate appropriate preventive measures. To reach this goal, may become difficult because many years after the emergence of the new virulent strain (027 type) in parallel with others, which are eventually antibiotic-resistant (e.g., moxifloxacin-resistant, ribotype 012, 017, and 046 isolated from Sweden) [17], some authors highlight the lack of consensus regarding case definition, sampling, and diagnosis step algorithms [18, 19].

2.1. Direct detection of *C. difficile* in clinical specimens by molecular platforms

Different protocol algorithms use two or three steps to improve diagnosis sensitivity and specificity, looking to the lowest cost as well. In two-step algorithm that includes as first test the GDH detection, the negative test proves the absence of *C. difficile*. For the positive test, the EIA and/or the nucleic acid amplification test (NAAT) can be used as confirmatory tests. In the two-step algorithm, the NAAT is used after the GDH detection, and the positive test will impose the patient's isolation and the preventive methods to stop the transmission. The two-step algorithm may fail in CDI diagnosis, the GDH detection being related to the test sensitivity (79–98%) (Shetty et al., cited by Tenover et al. [20]); low sensitivity is related to different nontype 027 strains [20].

For the three-step variant, the EIA is used as confirmatory test and for negative results, and the NAAT is necessary to confirm or note the negative results (**Figure 1**).

Using directly molecular methods, the *C. difficile* diagnosis rates become twofold higher in comparison with EIA alone. Soon after that, the incidence of *C. difficile* increased but decreased afterward, probably because the experts used new case definitions, and better preventive methods can explain the transmission decline in hospitals. After many years of using in parallel different diagnosis algorithms, there are no sufficient data to support a clear conclusion regarding the clinical benefit and hospital costs. Larson et al.'s study concludes that PCR alone can save around 200,000 USA dollars annually, mainly by removing several tests [21]. Burnham and Carroll clearly mentioned that the need of future cost-efficient studies relates to NAAT testing alone [22, 23].

Phenotypical methods	Toxin A/B tested by EIAs
	Glutamate dehydrogenase antigen (GDH)
	EIA screen—GDH/toxin
	Lateral flow, Membrane assay
	Cell culture cytotoxin neutralization assays (CCCNA)
	Toxigenic culture (Culture + Cytotoxin assay) "Gold standard"
Molecular methods	Molecular strains typing
	PCR amplification
	Sequence-based

Table 1. Main techniques for the detection of *C. difficile* [13].

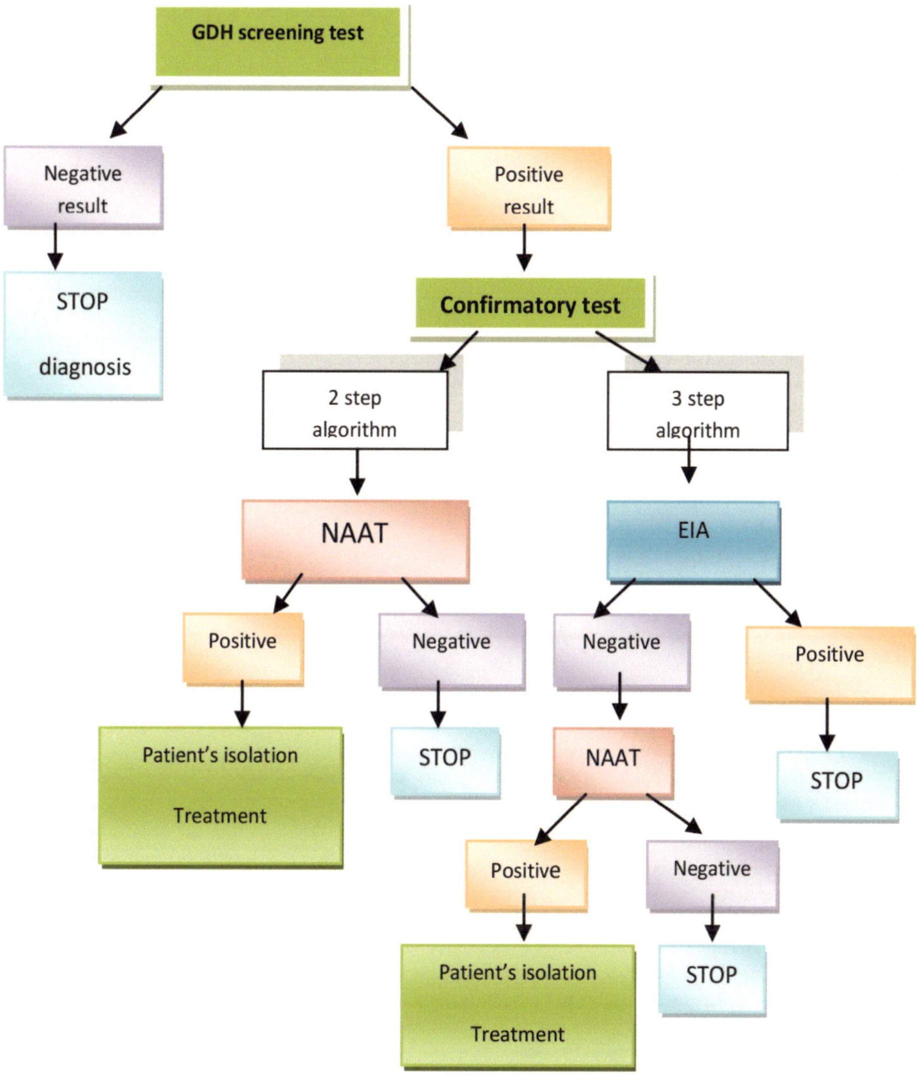

Figure 1. Algorithm steps for *C. difficile* diagnosis—different laboratory approaches.

3. Molecular methods applicable to *C. difficile* diagnosis

Until now, it has been clearly stated that molecular tests improve the detection of *C. difficile* in samples (stools, respectively), but that gene detection is not always a proof of their phenotypic expression. A few studies have tried to correlate gene detection with toxin expression using sequence analysis: some of them have found a major advantage for amplification methods in comparison with the EIA and GDH detection for the 027 isolates, but the small number of ribotypes belonging to the non-027 strains seems to be the main cause for test similarity results for this category [13] (**Table 2**).

Method	Comments	References
AP-PCR	Advantages: It is a reproducible, rapid, and simple technique It is sensitive and accurate, comparable with immunoblotting and REA, for the identification of *C. difficile* Applicability: In clinical laboratory, for strain identification Approximated time: 36–48 h; it is cost-efficient, especially for healthcare facility outbreak investigations Cost: $150–200 per specimen, when duplicate reactions with two sets of primers are used Disadvantages: Low reproducibility	[26–28, 36]
PCR ribotyping	Two sets of primers were proposed (USA *versus* UK) It is the reference standard method for *C. difficile* strain typing in Europe Advantages: Discriminatory power Applicability: Laboratory with moderate equipment Disadvantages: Moderate typability, reproducibility, ease of interpretation, and transportability	[10, 28, 33, 37, 54]
Repetitive extragenic palindromic sequence-based PCR (rep-PCR)	It uses heterogenous PCR primers that target noncoding repetitive sequences from the *C. difficile* genome Advantages: It is highly reproducible and more discriminatory than PCR ribotyping Applicability: *C. difficile* ribotyping Disadvantages: Even though the results for ribotypes 027 and 001 were excellent correlated with the PFGE and PCR ribotyping, for other types, it showed a reduced concordance. In the same studies, the method failed to separate between these two ribotypes Semiautomated variant (DiversiLab): Better standardization and reproducibility than the manual rep-PCR	[33, 37]
REA	It uses the *HindIII* enzyme in most of the protocols; a large number of bands can be separated by classical gel electrophoresis Applicability: It has been proved to be a very useful method for epidemiological studies, with an excellent discriminatory capacity and reproducibility Disadvantages: It is difficult to interpret band patterns, and protocols are difficult to compare between laboratories	[33, 35]
PFGE	The most frequent method used for *C. difficile* strain typing, as support for outbreak investigations Advantages: It is the "gold standard" method for *C. difficile* typing, as well as for other bacterium strains, and it is as comparative method between laboratories; it is a discriminatory and reproducible method, even though some strains are not typified In a modified protocol, the number of nontypeable strains has become 0, and both advantages were similar It is the reference standard in the USA Applicability: control of epidemic outbreaks Disadvantages: Needs longer time, it is more expensive, and it is designated for expertise laboratories because of the difficulties in the inter-laboratory comparison	[33–35, 37, 38]
Toxin-typing RFLP	It is based on the capacity of the *C. difficile* genome to encode a synthesis of minimum two toxins (A and B) (pathogenicity locus-PaLoc), and it is defined "as a group of strains with identical changes in the PaLoc when compared with the other strains," additional toxins (*tcd C, tcdR, tcdE*) and, for some strains, binary toxin too. There are 32 toxinotypes (some of them described as "minor," and 0 type) (from reference strain VPI10463) The current method for toxinotyping is restriction length polymorphism (RFLP) Advantages: Highly reproducible Applicability: In well-equipped laboratories for epidemiological studies. Toxinotypes III, IV, V, VIII, IX, and XII, the most frequent isolated from humans (VIII and IIIb [BI/NAP1/027 strains], are associated with disease severity and have been isolated worldwide) Disadvantages: Lack of consensus standards, difficult interpretation	[33, 39–42, 45]

Table 2. The main biological molecular methods for *C. difficile* infection diagnosis.

3.1. Amplification methods

For many years, the nucleic acid amplification tests (NAATs) have been used to detect *C. difficile* in fecal samples as an efficient tool to replace nonmolecular methods such as EIA and GDH; the first publication appeared in the early 1990s and the conventional PCR methods had as target the different genes (e.g., *tcdA, tcdB*, and *16S rRNA* genes, respectively) [13, 22].

The emergence of the new *C. difficile* virulent strains and the low sensitivity and specificity of the phenotypic methods have imposed their replacement with molecular methods, starting with genotyping and strain typing, respectively. In a very well-documented presentation of these methods, Dingle and MacCannel [24] proposed a list of typing methods. Briefly, these methods are restriction fragment methods (restriction endonuclease analysis [REA], pulsed field gel electrophoresis [PFGE], toxinotyping, PCR amplification methods [PCR ribotyping], repetitive extragenic palindromic PCR [rep-PCR], multilocus variable-number tandem repeat analysis [MLVA], arbitrarily primed PCR [AP-PCR], and sequence-based methods, including the whole genome sequencing [WGS], target sequencing approaches, multilocus sequence typing [MLST], surface layer protein A sequencing [slpA], tandem repeat sequence typing [TRST], single nucleotide polymorphism typing [SNPT], whole genome multilocus sequence typing, and the Kmer-based comparison).

From these methods, we are presenting the one with the highest clinical applicability, and for the most important of them, we have conducted a synthesis of recent scientific literature (PFGE, ribotyping, MLVA, and sequence-based methods). We are concluding this presentation of methods for *C. difficile* detection by presenting the FDA commercially approved tests, which have the advantage of accuracy, reproducibility, and clinical validation by comparing different laboratories from different countries.

Finally, we will list the main contribution of the molecular methods regarding the antibiotic resistance surveillance, the comparison between the biological molecular methods and a comparison between the classical rapid detection tests (EIA, DGH) versus the modern biological molecular methods.

3.1.1. The arbitrarily primed PCR or AP-PCR

It is a simple and rapid method to detect different bacteria outbreaks [25]. The AP-PCR belongs to the early stages of biological molecular era and uses a selection of primers that, at low annealing temperatures, produced a variety of bands. Details regarding the patent of this technique (*Arbitrarily primed polymerase chain reaction method for fingerprinting genomes*; United States Patent 6696277) can be found in (http://www.freepatentsonline.com/6696277.html) [26].

Advantages—AP-PCR provides a rapid and sensitive screen for the determination of clonal relationships among *C. difficile* strains [27]. The comparative studies using PCR ribotyping and AP-PCR from the cited authors (e.g., McMillin, 1992; Van Belkum, 1993, 1994) used previous study primers (Van Belkum, 1994) and demonstrated that for samples in which the DNA was severely affected and the RFLP was unable to recognize fragments, the AP-PCR was able to provide clear typing results [28].

Disadvantages—It is less reproducible than the PCR ribotyping and, for this reason, it is recommended as the best typing method, even though it has concordant results with the AP-PCR [27]. The AP/PCR and the RAPD (random amplified DNA, that uses short primers) were not accepted later as comparative interlaboratory methods because of their difficulties in control and standardization (Colliers et al., cited by Gürtler and Grando [29]).

3.1.2. Restriction endonuclease analysis (REA)

In one of the first articles (1987) that related to the REA, Wren and Tabaqchali used *HindIII* as cutting enzyme and concluded that analyzing the DNA with it is a very sensitive method for *C. difficile* strain differentiation and a tool for epidemiological studies. Other restriction enzymes, such as *BamHI, EcoRI, SalI,* and *SmaI,* were less useful in a clear separation of bands [30]. In a very well cited paper [31], two extraction methods were used, and the 206 REA types were classified into 75 groups; they concluded that the REA is a rapid, very sensitive, discriminatory, and reproducible method for *C. difficile* typing, and recommended its use in large epidemiological studies. Using this efficient tool, different authors have also identified outbreaks from North America in USA and in Canada, respectively [7, 32]. Using the biological molecular methods, different study groups have clearly noted that the main strain that is spread over the world (Europe, USA, etc.) is known as BI/NAPI/027 because of different methods used to identify it (e.g., type BI, using REA; "NAP1," from the NA, North America, and P, from pulse-field type 1, using as method the PFGE; and 027 by PCR-ribotyping) [7, 31]. The REA is a relatively simple method for analyzing the total genomic DNA and has been successfully applied to several bacterial species, including *C. difficile.* Among the major advantages, we can note its good discriminatory capacity, but the main concern is related to the numerous restriction fragments that are generated, especially fragments smaller than 11 kb [33]. It is difficult to compare the method between laboratories, and its high-level protocol technicity required a high-level expertise [24].

3.1.3. Pulse field gel electrophoresis (PFGE)

For years, the epidemiology and clinical case management were imposed on *C. difficile* strain typing. Even though the PFGE was one of the first used methods in some regions as North America, it is still considered as being a standard method. In short, the PFGE supposed DNA digestion with restriction enzymes (e.g., *SmaI*) and agarose gel electrophoresis for DNA fragment separation and characterization. After that, an electric field is being changed repeatedly, using three directions. The large DNA fragments that cannot be clearly separated by classical gel electrophoresis are separated by an electric field that changes directions many times in 12–24 h. In literature, this clearer band pattern is known as the "North-American pulsed-field," or as the NAP types. One pulsotype is defined as that with more than 80% similarity, in interpreting such bands, we must bear in mind the subjectivity of interpretation and the fact that, sometimes, we cannot differentiate between the studied strain and the reference strains that we used for comparison [13]. The PFGE was the method used by many researchers, some of them looking at the comparability degree, using different methods (e.g., MLVA, REA, and PFGE); for example, Killgore et al. find that the D value (discriminatory index score) for PFGE is 0.843; in the cited study, the D varied between 0.964 and 0.631, and they recommend as reference methods, the REA and the MLVA, for their discriminatory capacity [32]. Even though it is considered as a "gold standard," the method has changed the protocol over the years and many protocols have been suggested (Corkill et al., 2000; Herschleb et al., 2007) [34, 35].

From the PubMed database, using the next keyword—*PFGE C. difficile human infections*—we have selected 48 articles, published between 1994 and 2017. In time, the PFGE has been used by researchers in different clinical and laboratory applications.

Starting with the 1990s, many authors have compared the PFGE with other molecular methods used to analyze *C. difficile* [33, 36, 37]. Some of them have concluded that the PFGE had comparable discriminatory powers for epidemiologic typing of *C. difficile* isolates and that the ribotyping is appreciably less discriminatory [33], whereas others have found that the PCR-based methods were easier and quicker to perform, but their fingerprints were more difficult to interpret than those of the PFGE [36].

Pasanen et al. concluded that PFGE and classical PCR ribotyping can be used as reference methods in epidemiological studies to confirm the results obtained with other methods. Because of the discrepancies between studies (e.g., some authors recommend REA and MLVA [32] for *C. difficile* typing, while others recommend PFGE and conventional PCR [37]), it is obvious that further studies and long periods of time are needed in different large geographical areas for the results to be statistically significant.

Another research team has detected *C. difficile* in infections other than the very well-known antibiotic-associated diarrhea, like the infection of a prosthetic joint [38] in association with inflammatory bowel disease [39] in which, interestingly enough, the nonclonal distribution of distinct strains was further demonstrated by the PFGE genomic fingerprinting. A team from Ohio, USA, published two cases of *C. difficile* bacteremia, which is a rare event. In one of the cases, the bacteremia was caused by the North American pulsed field gel electrophoresis (PFGE) type 1 (NAP-1) strain [40].

In the natural history of *C. difficile* infection, it is important to establish if the infection is a singular or a recurrent one of relapse or re-infection [41, 42]. The same method (PFGE) has been used to identify host and bacterial factors associated with healthcare-associated acquisition of *C. difficile* infection and colonization. Among patients with healthcare-associated *C. difficile* infection and those with colonization, 62.7 and 36.1%, respectively, had the North American PFGE type 1 (NAP1) strain [43].

In 2006, Kuijper et al. [44] concluded that the increased virulence of C. *difficile* is probably related to the association with antibiotic resistance because the fluoroquinolone-resistant strains are also highly toxigenic; these strains belonging to the ribotype 027 (toxinotype III) have been isolated in the last years from many countries, from hospitals (e.g., in England and The Netherlands, from 75 and 16 hospitals, respectively), and from healthcare facilities (e.g., Belgium and France) [44]. After this ECDC report, other authors have detected the rapid spread of *C. difficile* NAP1/027 in Brazil [45], Korea [46], Hong Kong [47], Latin America (Chile) [48], and Romania [49].

If the previously mentioned authors have associated the *C. difficile* strain NAP1/027 with recent outbreaks in North America and Europe, characterized by more severe disease symptoms, higher mortality rates, and greater risk of relapse, Sirard et al. [51] studied, in a nonoutbreak situation, whether specific strains, such as NAP1/027, were associated with more severe disease symptoms, higher toxin production, and/or greater sporulation *in vitro*. Their results suggested that some NAPI/027 strains can be isolated from less severe cases, and even though those strains produce large toxin amounts, probably other factors explain better the patients' evolution, and that future studies are necessary to clarify this discrepancy [50].

3.1.4. PCR ribotyping

The strain typing method must be used for epidemiological reasons; in Europe, the ribotyping is the most used method in *C. difficile* identification. For amplification, this method uses one fragment from the most constant region, from 16S to 23S RNA genes, the so-called intergenic spacer. In this operon, there are many copies from the genome of *C. difficile* and, using a single primer pair, we can obtain different bands with sizes from 200 to 700 bp [13, 37, 52]. For their evidence, viewing and comparing, we can use the migration in agarose gel, with commercial kits or their analysis with a dedicated software [32, 36, 51, 52].

The definition of Gürtler and Grando [29] seems to briefly and completely describe one of the most useful method in *C. difficile* typing, first introduced by Gürtler (1993), when an original primer set, given to large fragments, was very difficult to differentiate in agarose gel electrophoresis; later, different authors have tried to overcome these difficulties, and, finally, Bidet et al. (1999) used the appropriate primers (located nearest to ITS1). When the gel electrophoresis was used to separate the amplified bands, their patterns were difficult to interpret. Despite these disadvantages, the low price, the high sensitivity, and specificity were listed as the main positive arguments for the simple gel electrophoresis [29].

One major disadvantage of this method is the difficulty to compare the results from different laboratories and the lack of unique software analysis. In order to make this method applicable in all laboratories and increase the reproducibility, the high-resolution capillary gel electrophoresis was called on [53]. To standardize the method in countries like the UK, different laboratories have started using the unique protocol [13], including the differentiation of some types (such as the 014 and the 020) that cannot be distinguished using the conventional agarose gel-based PCR ribotyping. To increase the discriminative capacity, the capillary gel electrophoresis can be used, including for the subtyping of the 014 type characterization [54], variant that also needs standardization.

CDI is a major issue of concern in Europe and USA. For surveillance studies, even though reporting the CDI is not mandatory in all EU countries, the ECDC, as CDC Atlanta advises, initiates large surveillance programs to decrease the incidence and severity of CDI [55].

In a comprehensive review, with the permission of Public Health England, the authors published the prevalence of *C. difficile* ribotypes, detected from 2007 to 2011. The decreasing order of ribotypes was 27, 001, 106, 015, 002, 078, 014, 005, 023, 016, 014/020, 020, 017, 026, 017, 026 [13]. Since then, many authors have published their results, specific for their countries: PCR—ribotype 018 (Italy) [57], ribotype 176 (the Czech Republic) [58], the first two *C. difficile* ribotype 027/ST1 isolates being identified in Beijing [59], the PCR-ribotype 176 in Prague [60], another four PCR-ribotypes (027, 033, 078, and 126) in Italy [61], ribotype 258 (Qatar) [62], the first Polish ribotype 027 [63], ribotype 126 in Southern Taiwan [64], and ribotype 244— Australia [57, 65]. The prevalence of CDI with different ribotypes worldwide underscores the importance of local surveillance in detecting and controlling *C. difficile* infection.

Few countries are performing follow-up studies on *C. difficile* ribotype's recurrence and relapses [66]. In Sweden, 29 patients were positive in at least one of the follow-up tests; 16 had the same ribotype in follow-up tests, i.e., relapse, and 13 patients with a different ribotype, i.e., re-infection [67]. In a recent study (USA) [68], conducted in a small study group, 25 patients respectively, 5 of them were colonized and only 3 were classified as recurrent cases (12%). All eight patients had risk factor healthcare exposure, and no other risk factors were identified (e.g., antibiotics or proton pump inhibitor treatment); neither of them was diagnosed with the 027 ribotype in the follow-up period. By comparison, Komar et al. (2016, UK) has monitored patients with *C. difficile* infection over a 2-year period, and they found that the epidemic *C. difficile* 027/ST1 caused the majority of infections during the sampling period [69].

Another application of ribotyping is testing the susceptibility of *C. difficile* to antibiotics, according to the known virulence of specific ribotypes [69, 70]. Moreover, a team from the UK has implemented the antimicrobial stewardship, which is a key component in the reduction of healthcare-associated infections, particularly of the *C. difficile* infection (CDI). They have successfully restricted the use of cephalosporins and, subsequently, of fluoroquinolones. From an endemically high level of >280 cases per year in 2007–2008, the number of CDIs decreased to 72 cases in 2011–2012 [71]. The same antimicrobial stewardship program was used in Austria, where the ribotype 027 is prevalent. The reduction of moxifloxacin use, combined with provided structured information on CDI, was associated with an immediate decrease of CDI rates in this large community teaching hospital [72].

3.1.5. Multilocus variable-number tandem repeat analysis (MLVA)

A different typing method had no high discriminatory capacity, and new techniques were in place to overcome this disadvantage bearing in mind that, in the epidemiological investigation of different outbreaks, it is essential to identify the source of infection/etiological agent and its connection with secondary cases. The MLVA based on the amplification of different size fragments that can be easily seen using the capillary gel electrophoresis and the automated analysis of amplified fragments is one of them [55, 56]. In this large study from Europe that included laboratories from different countries, the MLVA was used to trace the 027 ribotype. Using the MLVA's capacity to discriminate and to monitor the transmission events in the hospital or in healthcare facility settings, many research groups used the MLVA [13, 54, 55, 73–78]. Different studies have concluded that despite its high discriminatory capacity, the MLVA is not a standardized method and that, for future inter-laboratory comparability, further large studies must be run across the world, using an identical protocol.

In the recent years, numerous teams have used the MLVA in many approaches. A research group from the Netherlands described the clinical and the microbiological characteristics of CDI among hospitalized children, using the MLVA. They have concluded that the *C. difficile* PCR ribotype 265 was most prevalent in children, this strain being rarely found in other countries, except for Belgium. The MLVA showed genetic relatedness between three-fourths of pediatric and adult ribotype 265 strains, without a clear epidemiological link [79].

Using the same MLVA Assay, Krutova et al. have performed a survey in 18 hospitals in the Czech Republic. They have found the spread of two *C. difficile* PCR ribotypes within 18 hospitals, underlying the importance of standardizing the CDI testing protocols and implementing the mandatory CDI surveillance in the country [80].

Kullin et al. have found that identical MLVA types occurred in different wards over time and that several patients were infected with identical strains. The patient-to-patient transfer and the unique infection events might cause the predominance of ribotype 017 strains in the cohort. The multidrug-resistant strains are a potential reservoir for future infections [81].

Usui et al. have analyzed the *C. difficile* prevalence among piglets in Japan to clarify the infection origin and the extent of the associated risk by using molecular and microbiological methods for both swine and human clinical and foreign isolates; the MLVA was able to connect the European 078 ribotype source, spread in Japan by imported pigs [82].

The MLVA was also used to identify the first case of *C. difficile* RT027 infection in the Czech Republic (CZ), the patient having previously been hospitalized in Germany, prior to moving to CZ [83].

A team from the UK (Oxford) has run a comparison between the MLVA and the WGS, and they have found that both methods were very similar despite the fact that they have analyzed different parts of the bacterial genome. With improvements in the WGS technology, it is likely that the MLVA locus data will be available from the WGS in the near future [84].

Another application of the MLVA has been put into practice in the Netherlands, where the authors have encountered an outbreak because of these two types occurring simultaneously in a 980-bed teaching hospital. The clonal dissemination has been investigated by the MLVA that showed persistent clonal dissemination of types 017 and 027 despite the appropriate infection control measures [77].

Manzoor et al. have developed in the UK an eMLVA (extended) scheme, which provides insight into the genetic diversity of the *C. difficile* population at both global and cross-infection clusters in patient levels, with the possibility of replacing the PCR ribotyping. This eMLVA scheme could discriminate clinically significant clusters, while maintaining a good concordance with the PCR ribotyping. The typing schemes containing only seven loci showed, in contrast, poor association with the PCR ribotyping [85].

3.1.6. Multiplex nucleic acids test

Sometimes, the *C. difficile* strain diversity makes the classic PCR fail in identifying different types of isolates (some producing both toxins encoded by their genes, *tcdA, tcdB*, and some not, including binary toxin, mediated by the *cdtA* gene). The real-time PCR multiplex type has been put into practice by different researchers [86] or a real-time variant, fluorescence-based multiplex PCR, to simultaneously detect *tcdA* and *tcdB* genes in the patients' stools [87].

The *C. difficile* genome imposed the simultaneous detection of several genes: *tcdA, tcdB*, binary toxin, and Δ117 (single pair deletion at 117nt in *tcdC* gene), for the detection of the 027 ribotype (known as the epidemic strain 027/NAP1/PI, respectively). In the last decade, the FDA have approved or cleared many NAATs, including the multiplex type. Some of them are loop-mediated isothermal amplification assays—LAMP—(e.g., Illumigene, Meridian Bioscience, Inc., that detect the *tcdA* gene from the conserved region), helicase-based amplification (e.g., Portrait Cdiff Assay, Great Basin; AmpliVue Cdiff Assay—Quidel Corporation), or based on array technology (e.g., Verigene Cdiff Assay, Nanosphere, a multiplex one, that uses PCR-amplified DNA in a nanoparticle-based assay, able to detect the *tcdA* and *tcdB* genes, the binary toxin gene, and the Δ117) [88]

The Verigene *C. difficile* Nucleic Acid Assay has a high sensitivity (98.7%), and a relatively low specificity (87.5%) compared to direct cultivation as the gold standard method, and it is recommended to be used as a test for the 027 ribotype identification, having been able to detect the previously listed markers [88].

3.2. The main sequence-based method for *C. difficile*

3.2.1. Multilocus sequence typing

In the last years, the MLST has become one of the most accurate methods, especially for the identification of different pathogens, including *C. difficile,* using the internal fragments of seven genes. The higher accuracy is related to these allele fragments that have different lengths (from 300 to 600 bp) and are easy to arrange precisely with the support of the automated DNA sequencer, in both strands. For each housekeeping gene, the different sequence present in each species genome will be given as distinct allele, for each isolate [13, 89, 90]. MLST generates high-throughput sequence data that can be uploaded from laboratories worldwide to a common web database [55].

C. difficile is theoretically well suited to MLST, as the species are relatively genetically heterogeneous. In 2004, MLST was introduced to study the population structure and global epidemiology of *C. difficile* (*aroE, ddl, dutA, tpi, recA, gmk,* and *sodA*), in order to analyze a group of *C. difficile* isolates [13, 90, 91, 92]. A major advantage of sequence-based typing methods like MLST is the ease of interpretation of the generated data. The sequence data are unambiguous and, therefore, objective, highly reproducible and easily exchangeable between laboratories. A practical disadvantage of MLST remains the relatively high cost of sequencing multiple targets, which could partially explain why MLST has not replaced the conventional PCR ribotyping in many European laboratories [55].

The analysis of recent scientific literature has revealed the main clinical applications of MLST. It is very important to first analyze the spread of epidemic strains, including the hypervirulent ones. This has been done, e.g., by researchers from Latin America, who have found that, in Chile, the most prevalent subtype (near to 80%) is subtype 1, related to the hypervirulent strain NAP1/027/ST1. The MLST analysis was capable of describing a high similarity (73%, respectively) of this subtype with nine different other subtypes, characterized by a similar 117 bp deletion (in the *tcdC* gene) [49]. The spread of epidemic strains of *C. difficile* has been also studied on Czech isolates, using the MLST [80]. Kuwata et al. have claimed that in their first study from Japan, based on MLST, both toxigenic and nontoxigenic *C. difficile* strains showed high genetic variation and that drug resistance was more likely related to toxigenic strains [93]. Some authors have used the MLST and have found that the ICU-acquired toxigenic *C. difficile* was not linked to those detected on admission. The active screening for toxigenic *C. difficile* was not considered to be a resource-efficient measure in settings with a low prevalence of colonization [86]. Another application of the MLST usage was to evaluate the dissemination of clones in hospitals and breeding-farms or a contamination in the slaughter-house, and the probability of interspecies transmission [94].

3.2.2. *Whole genome sequencing (WGS)*

Whole genome sequencing (WGS) is another sequence-based assay besides MLST, which has emerged as a promising sequence-based technique since it allows the detection of variations between *C. difficile* strains by, for example, single nucleotide polymorphisms (SNPs) analysis [95]. A study performed in England, using the WGS technologies to identify SNPs, identified three sublineages of *C. difficile* RT017 circulating in London. Like the notorious RT027 lineage, which has caused global outbreaks of *C. difficile* infection since 2001, the lineage of toxin-defective RT017 strains appears to be continually evolving [96]. Even more, a team from Switzerland has developed a double locus sequence typing (DLST) scheme as a tool to analyze *C. difficile* isolates. The results of DLST were compared with the ones from MLST: DLST had a higher discriminatory power compared with MLST and successfully identified all isolates of the study. The main advantage of DLST is including the absence of DNA extraction (polymerase chain reaction PCR is performed on colonies), no specific instrumentation, the low cost, and the unambiguous definition of types [97].

4. The new epidemiology of *C. difficile* spread

In the past 20 years, the use of antibiotics and, in many countries, their overuse are the main causes of the extended spread of hyper virulent and multidrug-resistant strains of *C. difficile*, related to the 027 ribotype, as well as to many others. All methods try to rapidly identify the infected strain, but there are no solid arguments in differentiating between infection and colonization (positive *C. difficile* diagnosis, in the lack of symptoms). An excellent synthesis regarding colonization (carriage) and its role as a source of infection was published recently by Furuya-Kanamory et al.; the need of colonized patient identification is obvious: their prevalence is higher than in symptomatic cases; they can become a dangerous source of infection, and preventive methods must target them [98]. The cited study also made a vast synthesis of prevalence colonization rate over the world, which varied in large limits: 0–15% for healthy adults, almost 30% in colonized patients with nontoxigenic strains, and 0–51% in elderly residents of healthcare institutions [98].

Clinical and epidemiological data must be taken into consideration in this differentiation. The new rapid phenotypical and genotypical methods were a real support for the laboratory diagnosis that, in recent years, was able to rapidly identify the etiology of CDI; the need for rapid identification is related to the patients' treatment and to the preventive measures. The alarming increase of cases is related to the highly virulent strain circulation growth and to these new sensitive and specific methods that increased the rate of positive diagnosis. All these factors have led to declaring CDI a public health problem [13, 23, 99]. The prevalence of CDI increased especially since 2003 (in the Quebec study, from Canada, this was fourfold times higher than before) [23]. Later, new reports from North America and Europe [7, 55] have claimed the spread of ribotype 027 strains, highly virulent and resistant to fluoroquinolones, as the main cause of this increasing prevalence. If medical care-associated infections were, from the beginning, the main cause of these alarming phenomena, other studies have clearly demonstrated the common origin for about one quarter of the cases (Lessa et al., cited by Burnham and Carroll [13]).

5. Commercially available real-time PCR test

In the last two decades, the laboratories have focused on using commercial kits, which have many advantages when compared to "in house" methods. We have chosen the real-time PCR tests, approved by the FDA, and they seem to be the most commonly used around the world, due to their performances.

The commercially available, real-time PCR tests for *C. difficile* diagnosis include the BD GeneOhm Cdiff (BD Diagnostics; San Diego, CA, USA) (target *tcdB* gene), Prodesse ProGastro Cd (Gen-Probe Inc.; San Diego, CA, USA) (target *tcdB* gene), Xpert *C. difficile* (Cepheid; Sunnyvale, CA, USA) (target *tcdB* gene and binary toxin), and Illumigene *C. difficile* (Meridian Biosciences; Cincinnati, OH, USA) (target *tcdA* gene), Real-Time PCR tests. Compared to the Cytotoxicity Neutralization Assay Culture (CNAC) and/or to the toxigenic culture, the PCR assays have been reported with sensitivities and specificities ranging from 77 to 100% and 93 to 100%, respectively [20, 86]. These tests have been demonstrated as being similar or even more efficient when compared with CCNA for children stool samples [86].

The FDA approved a list of nucleic acid-based tests, which analyze variations in the sequence, structure, or the expression of the deoxyribonucleic acid (DNA) in order to diagnose infection with an identifiable pathogen, including *C. difficile*. (http://www.fda.gov/medicaldevices/productsandmedical procedures/invitrodiagnostics/ucm330711.htm) [96].

The consultation of PubMed has allowed us to select 114 articles: 3 articles that used IMDx *C. difficile* for Abbott m2000 test, 2 with BD Diagnostics BD MAX Cdiff Assay, 2 with Quidel Molecular Direct *C. difficile* Assay, 3 with Verigene *C. difficile* Nucleic acid Test, 4—Portrait Toxigenic *C. difficile* Assay, 3—Simplexa *C. difficile* Universal Direct Assay, 8—Xpert *C. difficile*/Epi, 27—Illumigene *C. difficile* Assay, 54—Xpert *C. difficile*, 3—ProGastro Cd Assay, and 5—BD GeneOhm C. diff Assay.

For this short presentation, we have chosen the latest references that offered the most eloquent data regarding sensitivity, specificity, and, if available, positive predictive value and negative predictive value (**Table 3**) [101–111].

One feature of these articles is that they compare different platforms and analyze samples from different medical centers, and this action favors the evaluation of the test's accuracy and reproducibility. The sensitivity and the rapidity of the NAATs are excellent and fast (e.g., 90 min using Portrait Toxigenic test, or 65 min GeneXpert for 80 samples, simultaneously) making them reliable methods for the direct detection of *tcdA* and/or *tcdB* in stool specimens, compared with the toxigenic culture [100, 101]. Some differences in the sensitivity of the NAATs may partly depend on the number of toxigenic *C. difficile* in stool specimens. Considering the rapidity and the high specificity of the real-time PCR assays compared to the toxigenic culture, they

References	Country	Method	Sensitivity	Specificity
[101]	Korea	Xpert	90.0%	92.9%
		Max	86.3%	89.3%
		IMDx	84.3%	94.4%
		Illumigene	82.4%	93.7%
[106]	Korea	Xpert	82.8%	98.8%
		BD MAX	81.6%	95.8%
		IMDx	62.1%	99.4%
[103]	New York, USA	BD Max Cdiff assay (Max)	96.9%	95%
		Abbott m2000 assay (IMDx)	92.8%	100%
[104]	Washington, Seattle	Verigene CDF	95.2%	99.4%
		Simplexa Universal Direct assay	87%	100%
		BD MAX	87%	98.8%
[107]	Heidelberg, Germany	BD MAX Cdiff assay	90.5%	97.9%
		Xpert *C. difficile* test	97.3%	97.9%
[108]	USA	Lyra assay on the SmartCycler II	82.1%	98.8%
		ABI 7500 Fast DX	96.9%	85.7%
		ABI QuantStudio DX	89.3%	99.0%
[109]	California, USA	Simplexa Universal Direct	98%	96%
		AmpliVue assays	100%	100%
[105]	Japan	Verigene CDF test	96.7%	97.4%
[102]	USA	Portrait toxigenic *Clostridium difficile* assay	98.2%	92.8%
[110]	California, USA	Early prototype core molecular mirroring nuclear magnetic resonance detection platform (Mentor-100)	88.4%	87.0%
[111]	Chicago, IL, USA	Wampole toxin A/B EIA	85.4%	90.9%
		CdQCC (GDH antigen)	95.8%	89.6%
		BD GeneOhm	100%	100 %
		Progastro Cd RT-PCR	93.8%	99.3%

Table 3. Characteristics of the FDA-approved tests for the detection of *Clostridium difficile*.

can be used as the first test method for *C. difficile* infection/colonization. However, additional efforts should focus on the discrimination between infection and colonization. According to the overall performance of these assays, these results support the routine use of the said platforms for the detection of the toxigenic *C. difficile* in the clinical laboratories, a fact that will have a positive impact on patient care. Moreover, the Verigene CDF test is a novel nucleic acid microarray that reliably detects both *C. difficile* toxins A and B in unformed stool (liquid) specimens and appears to adequately identify the ribotype 027 isolates [93, 104]. Another advantage is that sample processing is minimal, these tests having proved to be simple, cost-efficient, and with broad applicability to panel-based approaches, potentially simplifying the workflow [105].

Another application of these assays is the possible reporting of the DNA load of toxigenic *C. difficile* in the stool sample, which may represent a solution. Using the Xpert platform, most samples with discrepant results had *C. difficile* concentrations below the illumigene limit of detection. The significance of the low-level *C. difficile* detection needs to be further investigated [105]. The estimated cost avoidance provided by a more rapid molecular diagnosis can be outweighed by the cost of isolating and treating PCR-positive/cytotoxin-negative patients [103, 104]. The costs, the clinical consequences, and the impact on the nosocomial transmission of treating and/or isolating patients—positive for toxigenic *C. difficile* by PCR but negative for *in vivo* toxin production—worth further study. Diagnostic algorithms combining immunoassays and NAATs could also improve the specificity and reduce the global cost of this analysis [108–110].

In a recent study from 2014, Deak et al. [109] summarized the efficiency of two test cleared by the FDA (Simplexa Universal Direct and AmpliVue *C. difficile* Assay) having as references the Meridian Illuminigene Assay and the toxigenic *C. difficile* culture. In short, the best results were obtained with Simplexa (sensitivity 98%; limits: 88 ÷ 99.9), with short specimen preparation time, as AmpliVue (3 min, respectively), a small difference regarding the total time of the procedure (73 min *versus* 68 min for Illuminigene, and 91 min for AmpliVue), and only 8 min for hands-on time (*versus* 18, and 11 min for Illuminigene and AmpliVue, respectively). Both evaluated tests showed the same specificity (100%; limits: 96.9 ÷ 100). In conclusion, better parameters regarding the handling time, the higher sensitivity, and the possibility to use the Simplexa platform for other tests from Focus Diagnostics, recommend the kit for large laboratories with high sample numbers; the good performances of the AmpliVue can also be taken into consideration, when rapid and sensitive methods must be used for the *C. difficile* infection diagnosis [109].

6. Biological molecular methods for the characterization of *C. difficile* multidrug-resistant (MDR) strains

The antibiotic resistance explains CDI and changes the therapeutic scheme of these cases. New patterns of *C. difficile* strains encourage the spread and persistence of these strains in healthcare facility settings; for this reason, the surveillance of resistance and the molecular characterization of these mechanisms are very important in CDI control and prevention [111]. Both conjunction [112] and translocation of the genetic mobile element CTn5 [113] were described. The genetic mechanisms were much better understood and easier to study after 630 strain genome sequences [114] and resequencing studies. In a recent study, van Eijk et al. [114] analyzed the *C. difficile* laboratory strain 630Δerm and demonstrated, for the first time, the analysis of major methylation patterns for any *C. difficile* strain. In this research, the authors revealed that in addition to insertions, deletions, and SNPs, the CTn5 element moved from its original location within the CD1844 to the *rumA* gene, in its isolate. This molecular genetic study highlighted that the major rearrangement has important implications for the redistribution of strains with highly mobile genomes and that, even though it is not related to direct studies regarding the antibiotic resistance of *C. difficile,* it argues for the complete re-sequencing of common lab strains in each laboratory [113].

The complete genome sequences of *C. difficile* have made it possible to study the mobile genetic elements (transposons or transposable elements). This analysis describes that about 11% of the *C. difficile* genome consists of mobile genetic elements, the majority being conjugative transposons; their presence is at the origin of virulence, antibiotic resistance, synthesis of different surface protein, and microorganism-host regulation-binding capacities [114]. The horizontal genetic transfer of these transferable elements between species or between strains of *C. difficile* is the main cause of such successful spreading, especially in healthcare facilities, as well as within the community.

The interest in such studies started with the worldwide spread of the highly virulent RT027 and of other ribotypes, which explain the CDI treatment failure. Generally speaking, the resistance phenomena are the consequences of antibiotic treatment, especially clindamycin (CLI), cephalosporins (CFS), and fluoroquinolones (FQs), even though all antibiotics are theoretically capable to be the cause of CDI, in parallel with other risk factors (previously listed). The first antibiotics associated with CDI was CLI, a large broad spectrum antibiotics, with bacteriostatic activity, for which *C. difficile* demonstrated a high resistance level (more than 90%) [60]. The relative risk (RRs) for *C. difficile* acquisition is 9.0 for CLI, varying in large limits for CFs (from 7.8 for cefaclor, to 36.2 for cefotaxime), with lower risks for β-lactams (2.0 for penicillin, to 22.1 for ampicillin and amoxicillin-clavulanic acid association) [5]. Even though the FQs have been used since 1988, their implication in CDI was described in the last 10 years (Sunenschine et al., cited by Bartlett and Gerding [5]). The most studied strains that belong to the RT027 demonstrated a high-level resistance to the FQs [7]. An excellent synthesis regarding the *C. difficile* antibiotic resistance and associated RTs was recently published by Spigaglia, in 2016 [61]. Although for the antibiotic susceptibility studies, the most frequent methods are the agar dilution (AD) and the epsilometer test (Etest), ribotyping is a supportive method that classified the strains, to connect infection/microorganism's source/sources with secondary cases, in order to establish and implement the most appropriate prevention methods. Other methods can be used to study the genetic mechanisms of antibiotic resistance and their genetic horizontal transfer (by conjunction, or conjugative-like processes), or with the support of bacteriophages in transduction, including mobile genetic elements, as support for pathogenicity, antibiotic resistance, and environmental survival capacity [116–118]. Recent protocols were tested for the genetic manipulation of *C. difficile* genome and future researches will be able to continue this new approach [119].

The rate of antibiotic resistance varied between different studies depending, most probably, on the antibiotics policy and the spread of different *C. difficile* strains, in various regions. A large synthesis of 30 studies dating from 2012 to 2015 [115] revealed that resistance to CLI and CFs is higher than 50% (55 and 51%, respectively), with a similar percent for ERY and FQs (47%). One of the most used classes of antibiotics, the cephalosporins (CFs), is the source of many cases of CDI; the new CFs, with a wider antibiotic spectrum and more recently introduced in therapy, will induce the resistance phenomenon after several years of use. Thus, the second-generation cephalosporin (cefotetan/CTT and cefoxitin/FOX) level of resistance is about 80% and, for the third generation, recently introduced to the market (e.g., ceftriaxone/CRO and cefotaxime/CTX), the level of resistance has not yet reached 40% [115]. Although the phenomenon was highlighted by numerous strains resistant to CFs, the resistance mechanism is not completely understood. The strains belonging to different RTs (027 and to others ribotypes as it follows: 001, NAP1, 2, 4, and 6 [identified by PFGE], 002, 014, 017, 018, 053, 078, etc.) and *C. difficile* seem to be "constitutively resistant" to CFs, even though the main cases of CDI were described after CFs therapy, having different minimum inhibitory concentrations (MICs) to different CFs. These different MIC values are probably due to drug-resistant strains. The WGS 630 strain analysis shows 25 coding sequence (CSs); their presence is not constant in all strains, the percentage of identity ranging between 73 and 100% [115].

Not infrequently, the growing alarming statistic indicators (incidence, prevalence, etc.) are the starting point of some research that explains the epidemiological phenomenon using molecular biology techniques. Looking to CDI incidence dynamics, Clabots et al. explained the increased incidence of CDI cases in 1985, at the Minneapolis Veterans Administration Medical center (from 7.1 to 17.3/month), by highlighting a cryptic plasmid of 3.1 kilobases (kb), present exclusively in clindamycin-resistant strains and absent in sensitive isolates. The identification of this plasmid was carried out by restricting endonuclease digestion and Southern blot hybridization, but the authors failed in their purpose to identify the source of these strains and plasmid mechanism acquisition [120].

A large pan-European longitudinal surveillance of antibiotic resistance, conducted by Freeman J, et al. [122] enrolled around 1000 strains from 22 countries and 39 sites. Nine hundred and fifty-three strains were tested for PCR ribotyping, toxins, and antibiotic susceptibility for metronidazole (MTZ), vancomycin (VAN), fidaxomicin (FDX), rifampicin (RIF), moxifloxacin (MOX), clindamycin (CLI), imipenem (IM), chloramphenicol (CH), and tigecycline (TIG). CL, MOX, and RIF resistance levels varied (50, 40, and 13%, respectively) and were evident in many RTs. The most frequent RTs registered were 027 (12%), 001/072 (9%), 078, and 014 (8% each). MTZ, VAN, and TG were active in almost all cases, with a very low resistance level (2.18, 3.16, and 0.44%, respectively). One strain from UK (belonging to RT106) had a MIC for MTZ, 8 mg/L, and 20 strains demonstrated a reduced MTZ susceptibility, while 11 of them were RT 027. The reduced VAN susceptibility was very rare; four countries (the Czech Republic, Ireland, Latvia, and Poland) sent strains with MIC of 4 mg/L, Italy and Spain submitted different RTs with VAN reduce susceptibility, including resistant strains, belonging to RTs 027, 126, 356, and 001/007 [121]. The highest MIC values for VAN were determined for Rt 018 and 356.

Because the rates of treatment failure and recurrences have increased after MTZ and VAN cure, their replacement with different antibiotics was suggested. Rifamycins, like rifaximin (RFX) and fidaxomicin (FDX), a new bactericidal narrow spectrum macrocyclic antibiotic can be alternative therapies. Using antibiotic susceptibility test interpretative criteria, scientists proposed the use of rifampin (RIF) as an antibiotic related to RFX. The extrapolated literature data (from 2008 to 2012), from six relevant studies [115], show that 11% of *C. difficile* strains isolated from different clinical cases demonstrated resistance to RIF, mainly because this antibiotic is the drug of choice for tuberculosis treatment in many countries. All strains from Italy belonged to RT046, and the studied strains were isolated from TB patients with long RIF treatment [115]. In a study from Poland, from 2014, that characterized the pattern of susceptibility and ribotype association [121] for 83 strains, the majority belonged to RT027 (57.8%), 25.3% to RT176, and 16.9% to others. The majority of strains (85.5%) were resistant to erythromycin (ERY), more than a quarter percent (27.7%), resistant to CLI, with high MIC (greater than 256 mg/L). A high percent (83.1%) from all 83 strains was resistant to MOX, 87.9% resistant to imipenem (IMP), and only 2.4% to tetracycline. MTZ and VAN seem to continue to be efficient, having a low MIC_{90} value (0.75 mg/L) for both antibiotics. The most resistant strains belonged to RT027 and 176, with a MIC for ERT higher than 256 mg/L; the majority (95.2%) of RT176 strains were co-resistant to ERY and CLI. All strains resistant to ERY, MOX, and RIF belong to RT027 (18% strains respectively), and MDR strains (defined as strains resistant to at least three classes of antibiotics) were established for 71 strains (85.5%) [119] a percent that is an alarming phenomenon.

It is obvious that all articles from previous years that have studied the resistance patterns of *C. difficile* strains from clinical samples, animal, and environmental sources [108, 109, 111, 112, 120–131] from Europe, North America, and South-East Asia were designated not only to describe these patterns and their dynamics but also to classify them as RTs, to have a complete description of spread and risk for CDI, for higher virulent and MDR strains, too. Different studies have tried to describe, by different biological molecular methods, the *C. difficile* strain resistance mechanisms for cephalosporins, macrolide-lincosamide-streptogramin B (MLS_B) family, fluoroquinolones, including for antibiotics useful in the treatment of CDI t (e.g., metronidazole, vancomycin, rifamycins, and fidaxomicin); excellent syntheses regarding these mechanisms were recently published [61]. Having such a vast view on the evolution of antibiotic resistance to classical and new antibiotics, new strategies to limit this phenomenon could be designed and implemented to reduce hospitalization days, healthcare costs, and, not least, the emergence of new cases, potentially fatal.

7. Comparison of molecular biological methods

Nowadays, the revolution in the biological molecular field has generated a long list of methods and commercial kits, all these being found in numerous studies. Patient studies on different risk groups, reviews of literature, including meta-analyses and book chapters have been published in recent years, regarding the new phenomenon, CDI. The increasing incidence, the morbidity, and mortality by CDI in the world stimulate scientists to find the cause and solutions. Starting from years ago and even today, EIA and GDH for toxins and antigen detection were largely used but the lack of sensitivity of these methods imposed the search for new solutions. In the last 30 years, the PCR-based methods have found numerous practical applications for diagnosis, disease surveillance and, last but not least, for the study of antibiotic resistance, including for the highly virulent *C. difficile* strains. The main criteria to reassess variants of molecular biology tests are discriminatory power, reproducibility, technical difficulty, time needed to perform them, their cost-benefit ratio, the ease of interpreting the results, and data inter-comparison between laboratories [24]. Starting from such criteria, Dingle and MacCallenn established that the restriction fragment techniques (REA, PFGE, and toxinotyping) have very good reproducibility and discriminatory capacities, with less power for the toxinotyping, for the second criteria [33]. PFGE, which needs a longer time, is more expensive and it is designated for laboratory expertise because of the difficulties in inter-laboratory comparison [13, 33].

For years, different studies have concluded that REA and PGFE present a comparable discriminatory capacity, useful in epidemiological large studies regarding *C. difficile* typing isolates and that ribotyping has a less discriminatory capacity [33].

Second major laboratory methods are based on PCR amplification (ribotyping, REP-PCR, MLVA, and AP-PCR). From this list, ribotyping seems to fulfill all criteria and, close to it, MLVA, even though it needs a longer time [33].

In the future, sequence-based methods (MLST, slpA, TRST, and WGS) will probably become the key to answer difficult questions, such as strains trace, outbreaks sources, and so on, even though, currently, these are mainly dedicated to research laboratories since they are characterized by high technicity, high cost, and time, as well as by difficulties relating to the inter-comparison between laboratories. One question in the daily activity has to do with the selection of the appropriate genotyping method. To answer it, it is necessary to list their reference characteristics such as validity (to be applicable to all studied strains), discriminatory capacity (to be able to make the difference between unrelated strains), and reproducibility (this must be applicable between and within laboratories). Other criteria are related to difficulties and the step number of the method, rapidity, and to the cost-efficiency ratio. With this choice, we must take into consideration the purpose of genotyping: in the rapid tracking of local outbreaks, MLVA seems to be the most useful method and, in long epidemiological studies, MLST, PFGE, and WGS [35].

WGS will soon be the mandatory method to carry out quality multicenter studies, as ECDC has proposed minimum four major infections as follows: carbapeneme- resistant *Enterobacteriaceae, Neisseria gonorrhoeae, C. difficile,* and MRSA (methicillin-resistant *S. aureus*). According to the ECDC's experts, WGS "will improve the accuracy and effectiveness of disease surveillance, outbreak investigation and evaluation of prevention policies by enhanced assessment of disease and drug resistance transmission dynamics" [130]. Even though WGS can offer data (e.g., "in silico" design microrestriction profile for enzymes used in different techniques) for sequence-based and nonsequence-based genotyping methods, it remains a cumbersome test, expensive and dedicated only to high expertise laboratories [35].

8. Comparison between classical and modern methods

After the FDA has approved the standardized methods/tests, many authors have tried to find answers to different questions: Are the methods of molecular biology a better variant and more cost-effective than the phenotypic methods? In many countries, even with a good technical infrastructure and highly qualified personnel, EIA and GDH are the most useful methods for rapid CDI diagnosis. These tests and others, based on the phenotypic characterization (e.g., slide agglutination for serotyping, antibiotic susceptibility pattern), have some disadvantages such as a low reproducibility and the incapacity to differentiate between large numbers of isolates. Different study groups underline that EIA for A and B toxin detection has a low sensitivity and poor specificity [22, 131]. As we know, the efficiency of the test depends on the capacity of test itself to trace the smallest amounts of antigens in samples with higher specificity (these are dependent on the antibody clonality, on the binding power to EIA support, etc.) and on the difference in strain circulation in different geographical areas. For some EIA kits, sensitivity was about 15% for some ribotypes [127]. One very useful test is DGH that, in comparison with the toxigenic culture, showed a sensitivity similar to the real-time PCR for RT027, but the lowest sensitivity for non-027 infections; the sensitivity of GDH determination may be 70%, in comparison with the Gene-Xepert Assay [127]. GDH positive results must be confirmed with a different method (e.g., culture cytotoxicity assays, EIA, or NAATs) because this antigen can provide such results for both toxigenic and nontoxigenic strains.

Different authors, including professional society experts, recommend the two-step diagnosis, having GDH as a screening test and a follow-up test to confirm toxin presence. A recent meta-analysis (Shetty et al., cited by Carroll and Loeffelholz [23]) has concluded that GHD is highly sensitive, with high negative predictive value and that it has conducted to the best results in the two-step algorithms. To avoid this second test for toxin detection, a new immunochromatographic test (ICT) was proposed for both simultaneous detections. This ITC variant (Cdiff Quik Chek Complete; TechLab, Blacksburg, VA) has a good sensitivity for GDH, but a less good one for toxin detection (from 61 to 78%) and, for this reason, different laboratories use it in three-step diagnosis algorithm: GDH positive sample and toxin negative need to be retested by one NAAT.

On the other hand, different expert groups from Europe and USA [22, 131] strongly recommend replacing EIA with more sensitive assays, in order to quickly identify the *C. difficile* infection (or colonization) and to rapidly implement the appropriate preventive measures. Caroll and Loeffelholtz [23] support this approach having as arguments the final cost, which is the lowest for laboratories that use more sensitive tests, and the reduced healthcare associated infection, as consequence of rapid and efficient preventive methods.

Nowadays, in many laboratories, EIA and PCR-based methods used for *C. difficile* diagnosis replaced cultivation—the best way to study antibiotic susceptibility from pure culture—as source of future studies, including sequence analysis, but the toxigenic culture remains impractical, in relation to the time needed for end results.

As we have already briefly presented, the FDA has approved many NAATs, and different studies have concluded that such platforms have the highest sensitivity and that results are comparable between them. Contrary to this statement, we must note that the NAATs can detect conserved regions of A and/or B toxin genes, and not the toxin itself. For this reason, these tests cannot differentiate between infection and colonization and, moreover, they require complex infrastructure and highly qualified staff, both of which are more expensive, and that future researches are necessary for a clear conclusion regarding their utility for diagnosing the infection in children [23].

Even though the laboratory methods applicable for the identification of *C. difficile* are diverse, their various disadvantages have imposed the use of new techniques, including strain identification by MALDI-TOF (matrix-assisted laser desorption/ionization-time-flight mass spectrometry) [13, 132–135]. The method has been recently introduced in the clinical laboratory practice to identify the bacteria and to describe the resistance profile to antibiotics [13]. MALDI-TOF is based on the identification of some molecules, on calculating flight time in a vacuum tube, after the first sample was co-crystallized on a matrix and irradiated with laser. Flight duration allows the calculation of the weight/load ratio for the detected ions, generating a spectrum that will be compared with a database, to identify the bacterium in the sample. Whole bacteria, as well as brute extracts, may be used as samples [132, 133].

Identification is based on comparing the profile of some unique proteins present in the bacterial wall, thus becoming specific biomarkers, useful including in strain typing. Initially, Reil et al. have used the method to identify a limited number of ribotypes, including the highly virulent and the multidrug-resistant one (027 ribotype, respectively), based on the molecular weight of the selected proteins (between 2 and 20 kDa). Subsequently, Razarrdi and Akerlund have improved the method by selecting an initial wider range for the molecular weight of the selected proteins from 30 to 50 kDA followed by an analysis within the range used in the initial experiments (2–20 kDa), using ribotyping as a reference test by PCR. This way, for epidemiological purposes, typing *C. difficile* strains based on high molecular weight (HMW) might be combined with PCR ribotyping. In addition, by analyzing some surface proteins, involved in the attachment to specific receptors, MALDI-TOF would provide information related to the virulence of the infecting strain, including the immune response toward them and diseases severity [133].

By using MALDI-TOF, after obtaining the culture on chromogenic media, the diagnosis is more sensitive (from 94.7 to 100%) and more specific (from 79.7 to 100%), thus with significant statistical differences ($p < 0.001$). Furthermore, the technique allows the exclusion of false-positive results obtained by using the chromogenic media, hence the role of MALDI-TOF as confirmatory test. The rapid detection (10 min) and the very low cost make the test more useful as an identification test, compared with molecular biology methods (e.g., 16S rRNA gene sequencing), including the control of nosocomial infections associated with antibiotics therapy [134, 135].

To draw a general conclusion, we can state that the best algorithm starts with GDH as a screening test, a confirmation of toxin presence using a highly sensitive test, ideally a NAATs variant, bearing in mind that GDH efficiency is related to the strain types from different geographical areas [13, 20, 23].

9. Future directions

The prevention of CDI involves many factors, starting with the correct diagnosis, but we should also think about prevention (contact/isolation), antimicrobial stewardship, and fecal bacteriotherapy [136].

In a recent report of the ECDC, the importance of defining CDI, recurrent CDI cases, and CDI case origin were mentioned. Also, ESCMID recommended the following possible algorithms for the CDI diagnosis:

- Screening with NAAT, confirmation with toxin A/B EIA
- Screening with both GDH and A/B EIA toxin, optional confirmation with NAAT or toxigenic culture
- Screening with GDH EIA, confirmation with A/B EIA toxin, optional second confirmation with NAAT or toxigenic culture [137].

In recent years, quick methods—such as immunochromatography, which require less than 30 min, thus being useful as a screening test—have been proposed for the diagnosis of CDI; however, future clinical studies are necessary. New technologies—such as multicapillary column gas chromatography, a quick method but with sensitivity and specificity values less than 90%, a fact that requires the improvement of the method to increase accuracy—have been proposed. Another proposal aims at associating the use of selective media with the Fluorescent In Situ Hybridization (FISH) test, which, during the same day, can identify the strain by typing, determine the resistance profile, and detect toxins, therefore, having a higher clinical relevance, compared to other methods [135].

Like in other clinical situations, in which the diagnosis methods have switched to automatization, there is also a need for standardization of *C. difficile* detection. The clinical validation of the assays on many samples, to easily compare and to rely on the results from different countries, is very useful. The surveillance protocol established by the ECDC will improve the case management and the preventive measures [137–139].

Author details

Luminița Smaranda Iancu*, Andrei Florin Cârlan and Ramona Gabriela Ursu

*Address all correspondence to: luminita.iancu@umfiasi.ro

"Grigore T. Popa" University of Medicine and Pharmacy Iaşi, Iaşi, Romania

References

[1] Guide to Preventing *Clostridium difficile* Infections. Association for Professionals in Infection Control and Epidemiology USA. 2013. DOI: 1-933013-54-0. http://apic.org/Resource_/EliminationGuideForm/59397fc6-3f90-43d1-9325-e8be75d86888/File/2013CDiffFinal.pdf

[2] Davies KA, Longshaw CM, Davis GL, Bouza E, Barbut F, Barna Z, Delmee M, Fitzpatrick F, Ivanova K, Kuijper EJ, Macovei IS, Mentula S, Mastrantonio P, Von Muller L, Oleastro M, Petinaki E, Pituch H, Noren T, Novakova E, Nyč O, Rupnik M, Schmid D, Wilcox MH. Underdiagnosis of *Clostridium difficile* across Europe: The European, multicentre, prospective, biannual, point-prevalence study of *Clostridium difficile* infection in hospitalised patients with diarrhoea (EUCLID). Lancet Infectious Diseases. 2014;**14**:1208-1219. DOI: 10.1016/S1473-3099(14)70991-0

[3] Popescu GA, Szekely E, Codiţă I, Tălăpan D, Şerban R, Ruja G. Ghid de diagnostic, tratament şi prevenire a infecţiilor determinate de *Clostridium difficile*. 2016. DOI: 978-973-0-022285-2http://www.cnscbt.ro/index.php/ghiduri-si-protocoale/520-ghid-diagnostic-tratament-si-prevenire-clostridium-difficile/file

[4] Cheng JW, Xiao M, Kudinha T, Xu ZP, Sun LY, Hou X, Zhang L, Fan X, Kong F, Xu YC. The role of glutamate dehydrogenase (GDH) testing assay in the diagnosis of *Clostridium difficile* infections: A high sensitive screening test and an essential step in the proposed laboratory diagnosis workflow for developing countries like China. Plos one 10(12): e0144604. https://doi.org/10.1371/ journal. pone.0144604

[5] Bartlett JG, Gerding DN. Clinical recognition and diagnosis of *Clostridium difficile* infection. Clinical Infectious Diseases. 2008;**46**:S12–S18. DOI: https://doi.org/10.1086/521863

[6] Carter GP, Rood JI, Lyras D. The role of toxin A and toxin B in *Clostridium difficile*—Associated disease. Past and present perspective. Gut Microbes. 2010;**1**(1):58-64. DOI: 10.1038/nature07822

[7] McDonald LC, Killgore GE, Thompson A, Qwens RC, Kazakova SV, Sambol SP, Kohnson S, Gerding DA. An epidemic, toxin gene-variant strain of *Clostridium difficile*. The New England Journal of Medicine. 2005;**353**:2433-2441. DOI: 10.1056/NEJMoa051590

[8] Khan FY, Elzouki AN. *Clostridium difficile* infection: A review of the literature. Asian Pacific Journal of Tropical Medicine. 2014;**7**:6-13. DOI: 10.1016/S1995-7645(14)60197-8

[9] Daniel A, Rapose A. The evaluation of *Clostridium difficile* infection (CDI) in a community hospital. The Journal of Infection and Public Health. 2015;**8**:155-160. DOI: 10.1016/j.jiph.2014.08.002

[10] Center for Diseases Control and Prevention (CDC). Antibiotic Resistance. Threats in the United States. 2013, Atlanta 2013. Microorganisms with a Threat Level of Urgent. Clostridium difficile. p 51. https://www.cdc.gov/drugresistance/pdf/ar-threats-2013-508.pdf

[11] Collins DA, Elliott B, Riley TV. Molecular methods for detecting and typing of *Clostridium difficile*. Pathology. 2015;**47**:211-218. DOI: 10.1097/PAT.0000000000000238

[12] Goudarzi M, Seyedjavadi SS, Goudarzi H, Aghdam EM, Nazeri S. *Clostridium difficile* infection: Epidemiology, pathogenesis, risk factors and therapeutic options. Scientifica. 2014;**2014**:916826. DOI: 10.1155/2014/916826

[13] Burnham CA, Carroll KC. Diagnosis of *Clostridium difficile* infection: An ongoing conundrum for clinicians and for clinical laboratories. Clinical Microbiology Reviews. 2013;**26**(3):604-630. DOI: 10.1128/CMR.00016-13

[14] Karanika S, Paudel S, Zervou FN, Grigoras C, Zacharioudakis IM, Mylonakis E. Prevalence and clinical outcomes of *Clostridium difficile* infection in the intensive care unit: A systematic review and meta-analysis. Open Forum Infectious Diseases. 2015;**3**(1):ofv186. DOI: 10.1093/ofid/ofv186. eCollection 2016

[15] Kyne L, Hamel MB, Polavaram R, Kelly CP. Health care costs and mortality associated with nosocomial diarrhea due to *Clostridium difficile*. Clinical Infectious Diseases. 2002;**34**(3):346-353

[16] Schroeder LF, Robilotti E, Peterson LR, Banaei N, Dowdyd DW. Economic evaluation of laboratory testing strategies for hospital-associated *Clostridium difficile* infection. Journal of Clinical Microbiology. 2014;**52**(2):489-496. DOI: 10.1128/JCM.02777-13

[17] Akerlund T, Alefjord I, Dohnhammar U, Struew I, Norén T, Tegmark-Wisell K, The Swedish Study Group. Geographical clustering of cases of infection with moxifloxacin-resistant *Clostridium difficile* PCR ribotypes *012*, *017* and *046* in Sweden, 2008 and 2009. Euro Surveillance. 2011;**16**(10): 1-7. http://www.eurosurveillance.org/ViewArticle.aspx?ArticleId=19813

[18] Khanafer N, Barbut F, Eckert C, Perraud M, Demont C, Luxemburger C, Vanhems P. Factors predictive of severe *Clostridium difficile* infection depend on the definition used. Anaerobe. 2016;**37**:43-48. DOI: 10.1016/j.anaerobe.2015.08.002

[19] Korman TM. Diagnosis and management of *Clostridium difficile* infection. Seminars in Respiratory and Critical Care Medicine 2015;**36**(01):031-043. DOI: 10.1055/s-0034-1398741

[20] Tenover FC, Baron EJ, Peterson LR, Persing DH. Laboratory diagnosis of *Clostridium difficile* infection. The Journal of Molecular Diagnostics. 2011;**13**(6):573-582. DOI: 10.1016 /j.jmoldx.2011.06.001

[21] Larson AM, Fung AM, Fang FC. Evaluation of *tcdB* real-time PCR in a three step diagnostic algorithm for detection of toxigenic *Clostridium difficile*. Journal of Clinical Microbiology. 2010;**48**(1):124-130. DOI: 10.1128/JCM.00734-09

[22] Cohen SH, Gerding DN, Johnson S, Kelly CP, Loo VG, McDonald LC, Pepin J, Wilcox MH. Clinical practice guidelines for *Clostridium difficile* infection in adults: 2010 Update by the Society for Healthcare Epidemiology of America (SHEA) and the Infectious Diseases Society of America (IDSA). Infection Control and Hospital Epidemiology. 2010;**31**(5):431-455. DOI: 10.1086/651706

[23] Carroll KC, Loeffelholz M. Conventional versus molecular methods for the detection of *Clostridium difficile*. Journal of Clinical Microbiology. 2011;**49**:49-52. DOI: 10.1128/JCM.00700-11

[24] Dingle TC, MacCanell DR. Molecular strain typing and characterization of toxigenic *Clostridium difficile*, In: Sails A, Tang Y-W, editors. Current and Emerging Technologies for the Diagnosis of Microbial Infections. 2015;**42:**329-357. Methods in Microbiology. Elsevier, ISSN: 0580-9517. DOI: 10.1016/bs.mim.2015.07.001

[25] Lemann F, Chambon C, Barbut F, Gardin C, Brière J, Lamber-Zechovsky N, Branger C. Arbitrary primed PCR rules out *Clostridium difficile* cross-infection among patients in a haematology unit. The Journal of Hospital Infection. 1997;**35**(2):107-115

[26] Arbitrarily primed polymerase chain reaction method for fingerprinting genomes. United States Patent 6696277. Available from: http://www.freepatents online.com/6696277.html

[27] Martirosian G, Kuipers S, Verbrugh H, Van Belkum A, Meisel-Mikolajczyk F. PCR ribotyping and arbitrarily primed PCR for typing strains of *Clostridium difficile* from a Polish Maternity Hospital. Journal of Clinical Microbiology. 1995;**33**(8):2016-2021

[28] Sabat AJ, Budimir A, Nashev D, Sá-Leão R, Van Dijl JM, Laurent F, Grundmann H, Friedrich AW, on behalf of the ESCMID Study Group of Epidemiological Markers (ESGEM). Overview of molecular typing methods for outbreak detection and epidemiological surveillance. Euro Surveillance. 2013;**18**(4):20380

[29] Gürtler V, Grando D. Reprint of new opportunities for improved ribotyping of *C. difficile* clinical isolates by exploring their genomes. The Journal of Microbiological Methods. 2013;**95**(3):425-440. DOI: 10.1016/j.mimet.2013.09.009

[30] Wren B, Tabaqchali S. Restriction endonuclease DNA analysis of *Clostridium difficile*. Journal of Clinical Microbiology. 1987;**25**(12):2402-2404

[31] Clabots CR, Johnson S, Bettin KM, Mathie PA, Mulligan ME, Schaberg DR, Peterson LR, Gerding DN. Development of a rapid and efficient restriction endonuclease analysis typing system for *Clostridium difficile* and correlation with other typing systems. Journal of Clinical Microbiology. 1993;**31**(7):1870-1875

[32] Killgore G, Thompson A, Johnson S, Brazier J, Kuijper E, Pepin J, Frost EH, Savelkoul P, Nicholson B, Van den Berg RJ, Kato H, Sambol SP, Zukowski W, Woods C, Limbago B, Gerding DN, McDonald LC. Comparison of seven techniques for typing international epidemic strains of *Clostridium difficile*: Restriction endonuclease analysis, pulsed-field gel electrophoresis, PCR-ribotyping, multilocus sequence typing, multilocus variable-number tandem-repeat analysis, amplified fragment length polymorphism, and surface layer protein A gene sequence typing. Journal of Clinical Microbiology. 2008;**46**(2):431-437

[33] Kristjánsson M, Samore MH, Gerding DA, DeGirolami PC, Bettin KM, Karchmer AW, Arbeit RD. Comparison of restriction endonuclease analysis, ribotyping, and pulsed-field gel electrophoresis for molecular differentiation of *Clostridium difficile* strains. Journal of Clinical Microbiology. 1994;**32**(8):1963-1969. DOI: 0095-1137/94/$04.00+0

[34] Alonso R, Martin A, Pelaez T, Marin M, Rodriguez-Creixems M, Bouza E. An improved protocol for pulsed-field gel electrophoresis typing of *Clostridium difficile*. Journal of Medical Microbiology. 2005;**54**(Pt 2):155-157

[35] Li W, Raoult D, Fournier PE. Bacterial strain typing in the genomic era. FEMS Microbiology Ecology. 2009;**33**:892-916. DOI: 10.111/j.1574-6976.2009.00182.x

[36] Bidet P, Lalande V, Salauze B, Burghoffer B, Avesani V, Delmée M, Rossier A, Barbut F, Petit JC. Comparison of PCR-ribotyping, arbitrarily primed PCR, and pulsed-field gel electrophoresis for typing *Clostridium difficile*. Journal of Clinical Microbiology. 2000;**38**(7):2484-2487

[37] Pasanen T, Kotila SM, Horsma J, Virolainen A, Jalava J, Ibrahem S, Antikainen J, Mero S, Tarkka E, Vaara M, Tissari P. Comparison of repetitive extragenic palindromic sequence-based PCR with PCR ribotyping and pulsed-field gel electrophoresis in studying the clonality of *Clostridium difficile*. Clinical Microbiology and Infection. 2011;**17**(2):166-175. DOI: 10.1111/j.1469-0691.2010.03221.x

[38] McCarthy J, Stingemore N. *Clostridium difficile* infection of a prosthetic joint presenting 12 months after antibiotic-associated diarrhoea. Journal of Infection. 1999;**39**(1):94-96

[39] Clayton EM, Rea MC, Shanahan F, Quigley EM, Kiely B, Hill C, Ross RP. The vexed relationship between *Clostridium difficile* and inflammatory bowel disease: An assessment of carriage in an outpatient setting among patients in remission. American Journal of Gastroenterology. 2009;**104**(5):1162-1169. DOI: 10.1038/ajg.2009.4

[40] Hemminger J, Balada-Llasat JM, Raczkowski M, Buckosh M, Pancholi P. Two case reports of *Clostridium difficile* bacteremia, one with the epidemic NAP-1 strain. Infection. 2011;**39**(4):371-373. DOI: 10.1007/s15010-011-0115-7.

[41] Kato H, Kato N, Watanabe K, Ueno K, Sakata Y, Fujita K. *Relapses or reinfections: Analysis of a case of Clostridium difficile*-associated colitis by two typing systems. Current Microbiology. 1996;**33**(4):220-223

[42] Oka K, Osaki T, Hanawa T, Kurata S, Okazaki M, Manzoku T, Takahashi M, Tanaka M, Taguchi H, Watanabe T, Inamatsu T, Kamiya S. Molecular and microbiological characterization of *Clostridium difficile* isolates from single, relapse, and reinfection cases. Journal of Clinical Microbiology. 2012;**50**(3):915-921. DOI: 10.1128/JCM.05588-11

[43] Loo VG, Bourgault AM, Poirier L, Lamothe F, Michaud S, Turgeon N, Toye B, Beaudoin A, Frost EH, Gilca R, Brassard P, Dendukuri N, Béliveau C, Oughton M, Brukner I, Dascal A. Host and pathogen factors for *Clostridium difficile* infection and colonization. New England Journal of Medicine. 2011;**365**(18):1693-1703. DOI: 10.1056/NEJMoa1012413

[44] Kuijper EJ, Coignard B, Tüll P; ESCMID Study Group for *Clostridium difficile*; EU Member States; European Centre for Disease Prevention and Control. Emergence of *Clostridium difficile*-associated disease in North America and Europe. Clinical Microbiology and Infection. 2006;**12**(Suppl 6):2-18

[45] Balassiano IT, Miranda KR, Boente RF, Pauer H, Oliveira IC, Santos-Filho J, Amorim EL, Caniné GA, Souza CF, Gomes MZ, Ferreira EO, Brazier JS, Domingues RM. Characterization of *Clostridium difficile* strains isolated from immunosuppressed inpatients in a hospital in Rio de Janeiro, Brazil. Anaerobe. 2009;**15**(3):61-64. DOI: 10.1016/j.anaerobe.2008.12.007

[46] Kim H, Lee Y, Moon HW, Lim CS, Lee K, Chong Y. Emergence of *Clostridium difficile* ribotype *027* in Korea. Korean Journal of Laboratory Medicine. 2011 July;**31**(3):191-196. DOI: 10.3343/kjlm.2011.31.3.191

[47] Baldan R, Trovato A, Bianchini V, Biancardi A, Cichero P, Mazzotti M, Nizzero P, Moro M, Ossi C, Scarpellini P, Cirillo DM. *Clostridium difficile* PCR ribotype *018*, a successful epidemic genotype. Journal of Clinical Microbiology. 2015;**53**(8):2575-2580. DOI: 10.1128/JCM.00533-15

[48] Cheng VC, Yam WC, Chan JF, To KK, Ho PL, Yuen KY. *Clostridium difficile* ribotype *027* arrives in Hong Kong. The International Journal of Antimicrobial Agents. 2009;**34**(5):492-493. DOI: 10.1016/j.ijantimicag.2009.04.004

[49] Aguayo C, Flores R, Lévesque S, Araya P, Ulloa S, Hormazabal JC, Tognarelli J, Ibáñez D, Pidal P, Duery O, Olivares B, Fernández J. Rapid spread of *Clostridium difficile*

NAP1/027/ST1 in Chile confirms the emergence of the epidemic strain in Latin America. Epidemiology and Infection. 2015;**143**(14):3069-3073. DOI: 10.1017/S0950268815000023

[50] Popescu GA, Florea D, Rafila A. *Clostridium difficile* is emerging in Romania: A story of *027* ribotype and excessive antibiotic consumption. Journal of Gastrointestinal and Liver Diseases. 2014;**23**(3):342-343

[51] Sirard S, Valiquette L, Fortier LC. Lack of association between clinical outcome of *Clostridium difficile* infections, strain type, and virulence-associated phenotypes. Journal of Clinical Microbiology. 2011;**49**(12):4040-4046. DOI: 10.1128/JCM.05053-11

[52] Stubbs SL, Brazier JS, O'Neill GL, Duerden BI. PCR targeted to the *16S-23S rRNA* gene intergenic spacer region of *Clostridium difficile* and construction of a library consisting of 116 different PCR ribotypes. Journal of Clinical Microbiology. 1999;**37**(2):461-463

[53] Westblade LF, Chamberland RR, MacCannell D, Collins R, Dubberke ER, Dunne WM, Jr, Burnham CA. Development and evaluation of a novel, semiautomated *Clostridium difficile* typing platform. Journal of Clinical Microbiology. 2013;**51**(2):621-624. DOI: 10.1128/JCM.02627-12

[54] Indra A, Huhulescu S, Schneeweis M, Hasenberger P, Kernbichler S, Fiedler A, Wewalka G, Allerberger F, Kuijper EJ. Characterization of *Clostridium difficile* isolates using capillary gel electrophoresis-based PCR ribotyping. Journal of Medical Microbiology. 2008;**57**(Pt 11):1377-1382. DOI: 10.1099/jmm.0.47714-0

[55] Knetsch CW, Lawley TD, Hensgens MP, Corver J, Wilcox MW, Kuijper EJ. Current application and future perspectives of molecular typing methods to study *Clostridium difficile* infections. Euro Surveillance. 2013;**24**(4):20381

[56] European *Clostridium difficile* Infection Study Network (ECDISNET). Europe: Supporting capacity building for surveillance of *Clostridium difficile*. Leiden: ECDIS-NET.Version 1.2., 2013. Available from: http://www.ecdisnet.eu/ [Accessed 16 January 2017]

[57] Barbanti F, Spigaglia P. Characterization of *Clostridium difficile* PCR-ribotype *018*: A problematic emerging type. Anaerobe. 2016;**42**:123-129. DOI: 10.1016/j.anaerobe.2016.10.003

[58] Polivkova S, Krutova M, Petrlova K, Benes J, Nyc O. *Clostridium difficile* ribotype *176*—A predictor for high mortality and risk of nosocomial spread? Anaerobe. 2016;**40**:35-40. DOI: 10.1016/j.anaerobe.2016.05.002

[59] Cheng JW, Xiao M, Kudinha T, Xu ZP, Hou X, Sun LY, Zhang L, Fan X, Kong F, Xu YC. The first two *Clostridium difficile* ribotype *027/ST1* isolates identified in Beijing, China—An emerging problem or a neglected threat? Scientific Reports. 2016;**6**:18834. DOI: 10.1038/srep18834

[60] Drabek J, Nyc O, Krutova M, Stovicek J, Matejkova J, Keil R. Clinical features and characteristics of *Clostridium difficile* PCR-ribotype *176* infection: Results from a 1-year university hospital internal ward study. Annals of Clinical Microbiology and Antimicrobials. 2015;**14**:55. DOI: 10.1186/s12941-015-0114-0

[61] Spigaglia P, Barbanti F, Morandi M, Moro ML, Mastrantonio P. Diagnostic testing for *Clostridium difficile* in Italian microbiological laboratories. Anaerobe. 2016;**37**:29-33. DOI: 10.1016/j.anaerobe.2015.11.002

[62] Al-Thani AA, Hamdi WS, Al-Ansari NA, Doiphode SH, Wilson GJ. Polymerase chain reaction ribotyping of *Clostridium difficile* isolates in Qatar: A hospital-based study. BMC Infectious Diseases. 2014;**14**:502. DOI: 10.1186/1471-2334-14-502

[63] Lachowicz D, Szulencka G, Obuch-Woszczatyński P, Van Belkum A, Pituch H. First Polish outbreak of *Clostridium difficile* ribotype *027* infections among dialysis patients. European Journal of Clinical Microbiology & Infectious Diseases. 2015;**34**(1):63-67. DOI: 10.1007/s10096-014-2204-x

[64] Hung YP, Lin HJ, Tsai BY, Liu HC, Liu HC, Lee JC, Wu YH, Wilcox MH, Fawley WN, Hsueh PR, Tsai PJ, Ko WC. *Clostridium difficile* ribotype *126* in southern Taiwan: A cluster of three symptomatic cases. Anaerobe. 2014;**30**:188-192. DOI: 10.1016/j.anaerobe.2014.06.005

[65] Lim SK, Stuart RL, Mackin KE, Carter GP, Kotsanas D, Francis MJ, Easton M, Dimovski K, Elliott B, Riley TV, Hogg G, Paul E, Korman TM, Seemann T, Stinear TP, Lyras D, Jenkin GA. Emergence of a ribotype *244* strain of *Clostridium difficile* associated with severe disease and related to the epidemic ribotype *027* strain. Clinical Infectious Diseases. 2014;**58**(12):1723-1730. DOI: 10.1093/cid/ciu203

[66] Marín M, Martín A, Alcolea A, Iglesias C, Alcalá L, Peláez T, Sánchez-Somolinos M, Bouza E. First case of autochthonous *Clostridium difficile* PCR ribotype *027* detected in Spain. Enfermedades Infecciosas Y Microbiologia Clinica. 2014;**32**(6):355-358. DOI: 10.1016/j.eimc.2013.07.004

[67] Sandell S, Rashid MU, Jorup-Rönström C, Ellström K, Nord CE, Weintraub A. *Clostridium difficile* recurrences in Stockholm. Anaerobe. 2016;**38**:97-102. DOI: 10.1016/j.anaerobe.2016.01.005

[68] Thabit AK, Housman ST, Burnham CD, Nicolau DP. Association of healthcare exposure with acquisition of different *Clostridium difficile* strain types in patients with recurrent infection or colonization after clinical resolution of initial infection. The Journal of Hospital Infection. 2016;**92**(2):167-172. DOI: 10.1016/j.jhin.2015.11.009

[69] Kumar N, Miyajima F, He M, Roberts P, Swale A, Ellison L, Pickard D, Smith G, Molyneux R, Dougan G, Parkhill J, Wren BW, Parry CM, Pirmohamed M, Lawley TD. Genome-based infection tracking reveals dynamics of *Clostridium difficile* transmission and disease recurrence. Clinical Infectious Diseases. 2016;**62**(6):746-752. DOI: 10.1093/cid/civ1031

[70] Gerding DN, Hecht DW, Louie T, Nord CE, Talbot GH, Cornely OA, Buitrago M, Best E, Sambol S, Osmolski JR, Kracker H, Locher HH, Charef P, Wilcox M. Susceptibility of *Clostridium difficile* isolates from a Phase 2 clinical trial of cadazolid and vancomycin in *C. difficile* infection. The Journal of Antimicrobial Chemotherapy. 2016;**71**(1):213-219. DOI: 10.1093/jac/dkv300

[71] 71.Holt HM, Danielsen TK, Justesen US. Routine disc diffusion antimicrobial susceptibility testing of *Clostridium difficile* and association with PCR ribotype *027*. European Journal of Clinical Microbiology & Infectious Diseases. 2015;**34**(11):2243-2246. DOI: 10.1007/s10096-015-2475-x

[72] Sarma JB, Marshall B, Cleeve V, Tate D, Oswald T, Woolfrey S. Effects of fluoroquinolone restriction (from 2007 to 2012) on *Clostridium difficile* infections: Interrupted time-series analysis. The Journal of Hospital Infection. 2015;**91**(1):74-80. DOI: 10.1016/j.jhin.2015.05.013

[73] Wenisch JM, Equiluz-Bruck S, Fudel M, Reiter I, Schmid A, Singer E, Chott A. Decreasing *Clostridium difficile* infections by an antimicrobial stewardship program that reduces moxifloxacin use. Antimicrobial Agents and Chemotherapy. 2014;**58**(9):5079-5083. DOI: 10.1128/AAC.03006-14

[74] Van den Berg RJ, Schaap I, Templeton KE, Klaassen CH, Kuijper EJ. Typing and subtyping of *Clostridium difficile* isolates by using multiple-locus variable-number tandem-repeat analysis. Journal of Clinical Microbiology. 2007;**45**(3):1024-1028

[75] Marsh JW, O'Leary MM, Shutt KA, Pasculle AW, Johnson S, Gerding DN. Multilocus variable-number tandem-repeat analysis for investigation of *Clostridium difficile* transmission in Hospitals. Journal of Clinical Microbiology. 2006;**44**(7):2558-2566

[76] Eckert C, Vromman F, Halkovich A, Barbut F. Multilocus variable-number tandem repeat analysis: A helpful tool for subtyping French *Clostridium difficile* PCR ribotype *027* isolates. Journal of Medical Microbiology. 2011;**60**(Pt 8):1088-1094

[77] Goorhuis A, Debast SB, Dutilh JC, van Kinschot CM, Harmanus C, Cannegieter SC. Type-specific risk factors and outcome in an outbreak with 2 different *Clostridium difficile* types simultaneously in 1 hospital. Clinical Infectious Diseases. 2011;**53**(9):860-869. DOI: 10.1093/cid/cir549

[78] Fawley WN, Wilcox MH. *Clostridium difficile* ribotyping network for England and Northern Ireland. An enhanced DNA fingerprinting service to investigate potential *Clostridium difficile* infection case clusters sharing the same PCR ribotype. Journal of Clinical Microbiology. 2011;**49**(12):4333-4337

[79] Van Dorp SM, Smajlović E, Knetsch CW, Notermans DW, De Greeff SC, Kuijper EJ. Clinical and microbiological characteristics of *Clostridium difficile* infection among hospitalized children in the Netherlands. Clinical Infectious Diseases. 2017;**64**(2):192-198. DOI: 10.1093/cid/ciw699

[80] Krutova M, Matejkova J, Kuijper EJ, Drevinek P, Nyc O. Czech *Clostridium difficile* study group: *Clostridium difficile* PCR ribotypes *001* and *176*—The common denominator of *C. difficile* infection epidemiology in the Czech Republic, 2014. Euro Surveillance. 2016;**21**(29): 1-11. DOI: 10.2807/1560 7917.ES.2016.21.29.30296

[81] Kullin B, Brock T, Rajabally N, Anwar F, Vedantam G, Reid S, Abratt V. Characterisation of *Clostridium difficile* strains isolated from Groote Schuur Hospital, Cape Town, South

Africa. European Journal of Clinical Microbiology & Infectious Diseases. 2016;**35**(10):1709-1718. DOI: 10.1007/s10096-016-2717-6

[82] Usui M, Nanbu Y, Oka K, Takahashi M, Inamatsu T, Asai T, Kamiya S, Tamura Y. Genetic relatedness between Japanese and European isolates of *Clostridium difficile* originating from piglets and their risk associated with human health. Frontiers in Microbiology. 2014;**5**:513. DOI: 10.3389/fmicb.2014.00513. eCollection 2014

[83] Krutova M, Nyc O, Kuijper EJ, Geigerova L, Matejkova J, Bergerova T, Arvand M. A case of imported *Clostridium difficile* PCR-ribotype 027 infection within the Czech Republic which has a high prevalence of *C. difficile* ribotype 176. Anaerobe. 2014;**30**:153-155. DOI: 10.1016/j.anaerobe.2014.09.020

[84] Eyre DW, Fawley WN, Best EL, Griffiths D, Stoesser NE, Crook DW, Peto TE, Walker AS, Wilcox MH. Comparison of multilocus variable-number tandem-repeat analysis and whole-genome sequencing for investigation of *Clostridium difficile* transmission. Journal of Clinical Microbiology. 2013;**51**(12):4141-4149. DOI: 10.1128/JCM.01095-13

[85] Manzoor SE, Tanner HE, Marriott CL, Brazier JS, Hardy KJ, Platt S, Hawkey PM. Extended multilocus variable-number tandem-repeat analysis of *Clostridium difficile* correlates exactly with ribotyping and enables identification of hospital transmission. Journal of Clinical Microbiology. 2011;**49**(10):3523-3530. DOI: 10.1128/JCM.00546-11

[86] Kilic A, Alam MJ, Tisdel NL, Shah DN, Yapar M, Lasco TM, Garey KW. Multiplex real-time PCR method for simultaneous identification and toxigenic type characterization of *Clostridium difficile* from stool samples. Clinical Microbiology. 2015;**35**:306-313. DOI: 10.3343/alm.2015.35.3.306

[87] Belanger SD, Boissinot M, Clairoux N, Picard FJ, Bergeron MG. Rapid detection of *Clostridium difficile* in feces by real-time PCR. Journal of Clinical Microbiology. 2003;**41**:730-734. DOI: 10.1128/JCM.41.2.730-734.2003

[88] Carroll KC, Buchan BW, Tan S, Stamper PD, Riebe KM, Pancholi P, Kelly C, Rao A, Fader R, Cavagnolo R, Watson W, Goering RV, Trevino EA, Weissfeld AS. Multicenter evaluation of the *Verigene Clostridium difficile Nucleic Acid Assay*. Journal of Clinical Microbiology. 2013;**51**:4120-4125. DOI: 10.1128/JCM.01690-13

[89] Maiden MC. Multilocus sequence typing of bacteria. Annual Review of Microbiology. 2006;**60**:561-588

[90] Maiden MC, Bygraves JA, Feil E, Morelli G, Russell JE, Urwin R, Zhang Q, Zhou J, Zurth K, Caugant DA, Feavers IM, Achtman M, Spratt BG. Multilocus sequence typing: A portable approach to the identification of clones within populations of pathogenic microorganisms. Proceedings of the National Academy of Sciences of the United States of America. 1998;**95**(6):3140-3145

[91] Lemee L, Dhalluin A, Pestel-Caron M, Lemeland JF, Pons JL. Multilocus sequence typing analysis of human and animal *Clostridium difficile* isolates of various toxigenic types. Journal of Clinical Microbiology. 2004;**42**(6):2609-2617

[92] Lemee L, Bourgeois I, Ruffin E, Collignon A, Lemeland JF, Pons JL. Multilocus sequence analysis and comparative evolution of virulence-associated genes and housekeeping genes of *Clostridium difficile*. Microbiology. 2005;151(Pt 10):3171-3180

[93] Kuwata Y, Tanimoto S, Sawabe E, Shima M, Takahashi Y, Ushizawa H, Fujie T, Koike R, Tojo N, Kubota T, Saito R. Molecular epidemiology and antimicrobial susceptibility of *Clostridium difficile* isolated from a university teaching hospital in Japan. European Journal of Clinical Microbiology & Infectious Diseases. 2015;**34**(4):763-772. DOI: 10.1007/s10096-014-2290-9

[94] Zhang X, Wang X, Yang J, Liu X, Cai L, Zong Z. Colonization of toxigenic *Clostridium difficile* among ICU patients: A prospective study. BMC Infectious Diseases. 2016;**16**:397. DOI: 10.1186/s12879-016-1729-2

[95] Rodriguez C, Avesani V, Taminiau B, Van Broeck J, Brévers B, Delmée M, Daube G. Investigation of *Clostridium difficile* interspecies relatedness using multilocus sequence typing, multilocus variable-number tandem-repeat analysis and antimicrobial susceptibility testing. The Veterinary Journal. 2015;**206**(3):349-355. DOI: 10.1016/j.tvjl.2015.09.001

[96] Cairns MD, Preston MD, Lawley TD, Clark TG, Stabler RA, Wren BW. Genomic epidemiology of a protracted hospital outbreak caused by a toxin A-negative *Clostridium difficile* sublineage PCR ribotype *017* strain in London, England. Journal of Clinical Microbiology. 2015;**53**(10):3141-3147. DOI: 10.1128/JCM.00648

[97] Stojanov M, Magalhaes B, Terletsky V, Basset P, Prod'hom G, Greub G, Senn L, Blanc DS. Development and evaluation of double locus sequence typing for molecular epidemiological investigations of *Clostridium difficile*. European Journal of Clinical Microbiology & Infectious Diseases. 2016;**35**(2):175-181. DOI: 10.1007/s10096-015-2525-4

[98] Furoya-Kanamory L, Marquess J, Yakob L, Riley TV, Paterson DL, Foster NF, Huber CA, Clements CA. Asymptomatic *Clostridium difficile* colonization: Epidemiology and clinical implications. BMC Infectious Diseases. 2015;**15**;516-526. DOI: 10.1186/s12879-015-1258-4

[99] Vindigni SM, Surawicz CM. *C. difficile* infection: Changing epidemiology and management paradigms. Clinical and Translational Gastroenterology. 2015;**6**:e99. DOI:10.1038/ctg.2015.24

[100] U.S. Food and Drug Administration [Internet]. 2017. Available from: http://www.fda.gov/medicaldevices/productsandmedicalprocedures/invitrodiagnostics/ucm330711.htm [Accessed 16 January 2017]

[101] Shin BM, Yoo SM, Shin WC. Evaluation of Xpert *C. difficile*, BD MAX Cdiff, IMDx *C. difficile* for Abbott m2000, and Illumigene *C. difficile* assays for direct detection of toxigenic *Clostridium difficile* in stool specimens. The Annals of Laboratory Medicine. 2016;**36**(2):131-137. DOI: 10.3343/alm.2016.36.2.131

[102] Denys GA. Portrait toxigenic *Clostridium difficile* assay, an isothermal amplification assay detects toxigenic *C. difficile* in clinical stool specimens. Expert Review of Molecular Diagnostics. 2014;**14**(1):17-26. DOI: 10.1586/14737159.2014.864239

[103] Stellrecht KA, Espino AA, Maceira VP, Nattanmai SM, Butt SA, Wroblewski D, Hannett GE, Musser KA. Premarket evaluations of the IMDx *C. difficile* for Abbott m2000 Assay and the BD Max Cdiff Assay. Journal of Clinical Microbiology. 2014;**52**(5):1423-1428. DOI: 10.1128/JCM.03293-13

[104] Gilbreath JJ, Verma P, Abbott AN, Butler-Wu SM. Comparison of the Verigene *Clostridium difficile*, Simplexa *C. difficile* Universal Direct, BD MAX Cdiff, and Xpert *C. difficile* assays for the detection of toxigenic *C. difficile*. Diagnostic Microbiology and Infectious Disease. 2014;**80**(1):13-18. DOI: 10.1016/j.diagmicrobio.2014.06.001.v

[105] Tojo M, Nagamatsu M, Hayakawa K, Mezaki K, Kirikae T, Ohmagari N. Evaluation of an automated rapid diagnostic test for detection of *Clostridium difficile*. PLoS One. 2014;**9**(8):e106102. DOI: 10.1371/journal.pone.0106102. eCollection 2014

[106] Yoo J, Lee H, Park KG, Lee GD, Park YG, Park YJ. Evaluation of 3 automated real-time PCR (Xpert *C. difficile* Assay, BD MAX Cdiff, and IMDx *C. difficile* for Abbott m2000 Assay) for detecting *Clostridium difficile* toxin gene compared to toxigenic culture in stool specimens. Diagnostic Microbiology and Infectious Disease. 2015;**83**(1):7-10. DOI: 10.1016/j. diagmicrobio.2015.05.005

[107] Dalpke AH, Hofko M, Zorn M, Zimmermann S. Evaluation of the fully automated *BD MAX Cdiff* and *Xpert C. difficile* assays for direct detection of *Clostridium difficile* in stool specimens. Journal of Clinical Microbiology. 2013;**51**(6):1906-1908. DOI: 10.1128/JCM.00344-13

[108] Beck ET, Buchan BW, Riebe KM, Alkins BR, Pancholi P, Granato PA, Ledeboer NA. Multicenter evaluation of the Quidel Lyra Direct *C. difficile* nucleic acid amplification assay. Journal of Clinical Microbiology. 2014;**52**(6):1998-2002. DOI: 10.1128/JCM.03089-13

[109] Deak E, Miller SA, Humphries RM. Comparison of *Illumigene, Simplexa,* and *AmpliVue Clostridium difficile* molecular assays for diagnosis of *C. difficile* infection. Journal of Clinical Microbiology. 2014;**52**(3):960-963. DOI: 10.1128/JCM.02354-13

[110] Yang P, Hash S, Park K, Wong C, Doraisamy L, Petterson J, Petti CA, Ward PM, Lee SH, Menon S, She RC. Application of nuclear magnetic resonance to detect toxigenic *Clostridium difficile* from stool specimens: A proof of concept. The Journal of Molecular Diagnostics. 2017;**9**: 230-235. pii: S1525-1578(16)30258-6. DOI: 10.1016/j.jmoldx.2016.09.012

[111] Vasoo S, Stevens J, Portillo L, Barza R, Schejbal D, Wu MM, Chancey C, Singh K. Cost-effectiveness of a modified two-step algorithm using a combined glutamate dehydrogenase/toxin enzyme immunoassay and real-time PCR for the diagnosis of *Clostridium difficile* infection. Journal of Microbiology, Immunology and Infection. 2014;**47**(1):75-78. DOI: 10.1016/j.jmii.2012.07.008

[112] Solomon K, Fanning S, McDermott S, Murray S, Scott L, Martin A, Skally M, Burns K, Kuijper EJ, Fitzpatrick F, Fenelon L, Kyne L. PCR ribotype prevalence and molecular basis of macrolide–lincosamide–streptogramin B (MLSB) and fluoroquinolone

resistance in Irish clinical *Clostridium difficile* isolates. The Journal of Antimicrobial Chemotherapy. 2011;**66**:1976-1982. DOI: 10.1093/jac/dkr275

[113] Wasels W, Monot M, Spigaglia P, Barbanti F, Ma L, Bouchier C, Dupuy B, Mastrantonio P. Inter- and intraspecies transfer of a *Clostridium difficile* conjugative transposon conferring resistance to MLSB. Microbial Drug Resistance. 2014;**20**:555-560. DOI: 10.1089/mdr.2014.0015

[114] Van Eijk E, Anvar SY, Browne HP, Leung WY, Frank J, Schmitz AM, Roberts AP, Smits WK. Complete genome sequence of the *Clostridium difficile* laboratory strain *630Δerm* reveals differences from strain *630*, including translocation of the mobile element *CTn5*. BMC Genomics. 2015;**16**:31. DOI: 10.1186/s12864-015-1252-7

[115] Sebaihia M, Wren BW, Mullany P. The multidrug-resistant human pathogen *Clostridium difficile* has a highly mobile, mosaic genome. Nature Genetics. 2006;**38**:779-786. DOI: 10.1038/ng1830

[116] Spigaglia P. Recent advances in the understanding of antibiotic resistance in *Clostridium difficile* infection. Therapeutic Advances in Infectious Disease. 2016;**3**:23-42. DOI: 10.1177/2049936115622891

[117] Smith CJ, Markowitz SM, Macrina F. Transferable tetracycline resistance in *Clostridium difficile*. Antimicrobial Agents and Chemotherapy. 1981;**19**(6):997-1003. DOI: 0066-4804/81/060997-07$02.00/0

[118] Mullany P, Allan E, Roberts AP. Mobile genetic elements in *Clostridium difficile* and their role in genome Function. Research in Microbiology. 2015;**166**(4):361-367. DOI: 10.1016/j.resmic.2014.12.005

[119] Roberts AP, Allan E, Mullany P. The impact of horizontal gene transfer on the biology of *Clostridium difficile*. Advances in Microbial Physiology. 2014;**65**:63-82. DOI: 10.1016/bs.ampbs.2014.08.002

[120] Bouillaut L, McBride SM, Sorg JA. Genetic manipulation of *Clostridium difficile*. Current Protocols in Microbiology. 2011;**Chapter 9**:Unit 9A.2. DOI: 10.1002/9780471729259.mc09a02s20

[121] Clabots CR, Peterson LR, Gerding DN. Characterization of a nosocomial *Clostridium difficile* outbreak by using plasmid profile typing and clindamycin susceptibility testing. Journal of Infectious Diseases. 1988;**158**(4):731-736

[122] Freeman J, Vernon J, Morris K, Nicholson S, Todhunter S, Longshaw C, Wilcox MH. Pan-European longitudinal surveillance of antibiotic resistance among prevalent *Clostridium difficile* ribotypes. Clinical Microbiology and Infection. 2015;**21**:248-248. DOI: /10.1016/j.cmi.2014.09.017

[123] Lachowicz D, Pituch H, Obuch-Woszczatynski P. Antimicrobial susceptibility patterns of *Clostridium difficile* strains belonging to different polymerase chain reaction ribotypes isolated in Poland in 2012. Anaerobe. 2015;**31**:37-41. DOI: 10.1016/j.anaerobe.2014.09.004

[124] Drudy D, Goorhuis B, Bakker D, Kyne L, Van der Berg R. Clindamycin resistant clone of *C. difficile* PCR ribotype *027* in Europe. Emerging Infectious Diseases. 2008;**14**:1485-1487. DOI: 0.3201/eid1409.071346

[125] Baines SD, Wilcox MH. Antimicrobial resistance and reduced susceptibility in *Clostridium difficile*: Potential consequences for induction, treatment, and recurrence of *C. difficile* infection. Antibiotics. 2015;**4**:267-298. DOI: 10.3390/antibiotics4030267

[126] Ackermann G, Tang Y J, Kueper R, Heisig P, Rodloff A, Silva J, Cohen SH. Resistance to moxifloxacin in toxigenic *Clostridium difficile* isolates Is associated with mutations in *gyrA*. Antimicrobial Agents and Chemotherapy. 2001;**45**(8):2348-2353

[127] Costa MC, Stampfli HR, Arroyo LG, Pearl DL, Weese JS. Epidemiology of *Clostridium difficile* on a veal farm: Prevalence, molecular characterization and tetracycline resistance. Veterinary Microbiology. 2011;**152**(3-4):379-384. DOI: 10.1016/j. vetmic.2011.05.014

[128] Spigaglia P, Barbanti F, Dionisi AM, Mastrantonio P. *Clostridium difficile* isolates resistant to fluoroquinolones in Italy: Emergence of PCR ribotype *018*. Journal of Clinical Microbiology. 2010;**48**:2892-2896. DOI: 10.1128/JCM.02482-09

[129] Spigaglia P, Barbanti F, Mastrantonio P. Multidrug resistance in European *Clostridium difficile* clinical isolates. The Journal of Antimicrobial Chemotherapy. 2011;**66**:2227-2234. DOI: 10.1093/jac/dkr292

[130] Tenover FC, Novak-Weekley S, Woods CW, Peterson LR, Davis T, Schreckenberger P, Fang FC, Dascal A, Gerding DN, Nomura JH, Goering RV, Akerlund T, Weissfeld AS, Baron EJ, Wong E, Marlowe EM, Whitmore J, Persing DH. Impact of strain type on detection of toxigenic *Clostridium difficile*: Comparison of molecular diagnostic and enzyme immunoassay approaches. Journal of Clinical Microbiology. 2010;**48**:3719-3724. DOI: 10.1128/JCM.00427-10

[131] Tenover FC, Tickler IA, Persing DH. Antimicrobial-resistant strains of *Clostridium difficile* from North America. Antimicrobial Agents and Chemotherapy. 2012;56:2929-2932. DOI: 10.1128/AAC.00220-12

[132] Pulidot MR, Garçia-Quitanillat M, Martin-Peňa R, Cisneros JM, McConnell J. Progress on the development of rapid methods for antimicrobial susceptibility testing. The Journal of Antimicrobial Chemotherapy, 2013;**68**:2710-2717. DOI: 10.1093/jac/dkt253

[133] Rizzardi K, Åkerlund T., High molecular weight typing with MALDI-TOF MS—A novel method for rapid typing of *Clostridium difficile*. PLoS One. 2015 Apr 29;**10**(4):e0122457. DOI: 10.1371/journal.pone.0122457. eCollection 2015

[134] Chen JH, Cheng VC, Wong OY, Wong SC, So SY, Yam WC, Yuen KY. The importance of matrix-assisted laser desorption ionization-time of flight mass spectrometry for correct identification of *Clostridium difficile* isolated from chromID *C. difficile* chromogenic agar. Journal of Microbiology, Immunology and Infection (2016), http://dx.doi.org/10.1016/j.jmii.2015.12.002

[135] Plüddemann A, Onakpoya I, Harrison S, Shinkins B, Tompson A, Davis R, Price CP, Heneghan C. Position Paper on Anti-Microbial Resistance Diagnostics. University of Oxford: Centre for Evidence-Based Medicine; 2015. Available from: http://www.cebm.net/wp-content/uploads/2015/07/AMR-Diagnostic-technologies_10-June-2015.pdf

[136] Tian TT, Zhao JH, Yang J, Qiang CX, Li ZR, Chen J, Xu KY, Ciu QQ, Li RX. Molecular characterization of *Clostridium difficile* Isolates from human subjects and the environment. PLoS One. 2016;**11**(3): e0151964. DOI: 10.1371/journal. pone.0151964

[137] CDC Technical document European surveillance of *Clostridium difficile* infections Surveillance protocol version 2.2 [Internet]. 2017. Available from: http: //ecdc.europa.eu/en/publications/publications/european-surveillance-clostridium-difficile.v2final.pdf [Accessed 17 January 2017]

[138] Sartelli M, Malangoni MA, Abu-Zidan FM. WSES guidelines for management of *Clostridium difficile* infection in surgical patients. World Journal of Emergency Surgery. 2015;**10**:38. DOI: 10.1186/s13017-015-0033-6

[139] Crobach MJ, Planche T, Eckert C, Barbut F, Terveer EM, Dekkers OM, Wilcox MH, Kuijper EJ. European society of clinical microbiology and infectious diseases: Update of the diagnostic guidance document for *Clostridium difficile* infection. Clinical Microbiology and Infection. 2016;**4**:63-81. DOI: 10.1016/j.cmi.2016.03.010

Chapter 5

Assays for Measuring *C. difficile* Toxin Activity and Inhibition in Mammalian Cells

Mary Ann Cox, Lorraine D. Hernandez, Pulkit Gupta,
Zuo Zhang, Fred Racine and Alex G. Therien

Additional information is available at the end of the chapter

http://dx.doi.org/10.5772/68127

Abstract

Clostridium difficile infections (CDIs) are the leading cause of hospital-acquired infectious diarrhea. The symptoms of CDI are caused by two exotoxins, TcdA and TcdB, which are structurally and functionally highly homologous. Both toxins bind to specific receptors on mammalian cells, are internalized through endocytosis, translocate to the cytoplasm, and inactivate Rho-type GTPases via covalent glucosylation. This leads to downstream events that include morphological changes and disruption of epithelial tight junctions, release of pro-inflammatory mediators, and cell death. Assays used to assess the effects of toxins on cells have historically relied on evaluation of cell rounding or quantitation of ATP levels to estimate cell death—assays which can be qualitative and variable. In this chapter, several assays are described that robustly and quantitatively measure early and late toxin-dependent events in cells, including (i) toxin binding, (ii) Rac1 glucosylation, (iii) changes in cellular morphology (measured as dynamic mass redistribution), (iv) loss of epithelial integrity (measured as transepithelial electrical resistance), and (v) cell death (measured as total cellular protein using a colorimetric assay). The assays were validated using the highly specific monoclonal antitoxin antibodies, actoxumab and bezlotoxumab, which neutralize TcdA and TcdB, respectively.

Keywords: *C. difficile*, toxins, cell-based assays, epithelial cells, antitoxins

1. Introduction

Clostridium difficile (*C. difficile*) is an anaerobic, gram-positive, spore-forming bacterium that colonizes the lower intestinal tract of patients whose normal gut microflora has been disrupted by treatment with broad-spectrum antibiotics [1]. The symptoms of *C. difficile*

infection (CDI)—which include diarrhea and, in severe cases, pseudomembranous colitis, colonic rupture, and death [1, 2]—are caused by two exotoxins, toxin A (TcdA) and toxin B (TcdB) [3]. Both toxins have similar structural and functional characteristics. After binding to specific receptors on the surface of gut epithelial cells, they are internalized through endocytosis, translocate to the cytoplasm, and inactivate Rho-type GTPases via covalent glucosylation [4–7]. This leads to a variety of downstream events, including morphological changes associated with disruption of epithelial tight junctions, release of pro-inflammatory mediators (including interleukin-1β, tumor necrosis factor alpha, and interleukin-8), and eventually cell death [3, 8]. The damaging effects on the gut epithelium and initiation of a host inflammatory response are thought to underlie the clinical manifestation of CDI.

Current treatment for *C. difficile* infections includes discontinuing the offending broad-spectrum antibiotic and initiating therapy with narrower spectrum agents such as vancomycin, metronidazole, or fidaxomicin [9, 10]. Unfortunately, these treatments do not directly address the damaging effects of the toxins on the gut and perpetuate the gut dysbiosis that caused CDI in the first place. As a result, up to 25% or more patients successfully cured of an initial episode of CDI with these antibiotics suffer a recurrent episode within days to weeks. To address this, recent approaches to CDI treatment, including vaccines and monoclonal antibodies, have focused on neutralizing the effects of TcdA and TcdB, specifically, rather than the organisms itself [11–13]. Foremost among these novel therapies is bezlotoxumab, the anti-TcdB antibody recently approved by the Food and Drug Administration for reducing recurrence of CDI in patients 18 years of age or older who are receiving antibacterial drug treatment of CDI and are at a high risk for CDI recurrence.

The renewed interest in toxin-directed therapies underscores the importance of having robust quantitative assays in place to assess the activity of the *C. difficile* toxins. Historically, studying the effects of TcdA and TcdB on mammalian cells has been hampered by time-consuming and subjective assays that rely, for example, on visualization of cells to assess cell rounding or on the variable quantitation of ATP levels to measure cell death [13]. Thus, there is a scarcity of robust quantitative assays that measure the various cellular events associated with the intoxication cascade, making it difficult to evaluate new toxin-directed agents. In this chapter, we describe multiple quantitative cell-based assays that were newly developed, or adapted and optimized from previous reports, and used to interrogate the effect of the *C. difficile* toxins on epithelial cells. The assays are validated using the highly specific and potent antitoxin antibodies, actoxumab and bezlotoxumab, which bind to and neutralize TcdA and TcdB, respectively [13–15].

2. Materials and methods

2.1. TcdA- and TcdB-binding assay (Western blot)

TcdA (1 μg/ml) or TcdB (0.1 μg/ml) (The Native Antigen Company, Upper Heyford, the UK and tgcBIOMICS, Bingen, Germany) was incubated with or without 200 μg/ml actoxumab or bezlotoxumab in Vero cell culture medium (Eagle's minimal essential medium (EMEM) supplemented with 10% fetal bovine serum (FBS), 100 U/ml penicillin, and 100 U/ml streptomycin)

http://dx.doi.org/10.5772/68127

for 30 min at 37°C; these mixtures were then chilled on ice and added to plates of pre-chilled Vero cells (ATCC, Rockville, MD). Plates were incubated for 30 min on ice to allow binding of toxins. Following incubation, plates were washed three times with cold phosphate buffered saline (PBS) and cells were harvested by scraping. Cell membranes were isolated at 4°C with the Mem-PER Plus Membrane Protein Extraction Kit (Thermo Scientific, Grand Island, NY), according to the manufacturer's instructions, and solubilized in a total volume of 100 μL solubilization buffer with HALT protease/phosphatase inhibitors (Thermo Scientific). Following addition of Laemmli sample buffer, samples were incubated for 5 min at 95°C and resolved by SDS PAGE in 4–12% polyacrylamide gels and transferred to a nitrocellulose membrane. The nitrocellulose membrane containing transferred protein was blocked in Odyssey blocking buffer (Li-Cor) followed by incubation with actoxumab, bezlotoxumab, or an anti-cadherin antibody (Cell Signaling Technology, Beverly, MA) as the primary antibody for 1 h at room temperature (RT). After washing, the nitrocellulose membrane was incubated with a goat anti-human IgG antibody coupled to IRDye® 800CW (Li-Cor) for 30 min at RT. After additional washing, bands were visualized using the Odyssey imaging system (Li-Cor).

2.2. TcdA-binding assay (flow cytometry)

TcdA, from ribotype 087 (The Native Antigen Company, Upper Heyford, the UK), was fluorescently labeled using the Lightning Link Atto488 Antibody Labeling kit (Novus Biosciences, Littleton, CO) as directed by the manufacturer. About 50 μg of lyophilized TcdA was reconstituted for a minimum of 30 min in sterile ddH2O at RT before adding the LL-modifier buffer. The toxin/LL-modifier buffer solution was transferred to a vial containing the lyophilized Lightning Link mix. The mixture was pipetted up and down and incubated at RT in the dark. After 5 h, LL-quencher buffer was added and incubated at RT in the dark for 30 min and then stored at 4°C until use the following day. Several concentrations of TcdA-Atto488 were incubated with or without 200 μg/ml actoxumab at RT for 60 min, protected from light. Samples were then chilled on ice. Adherent HT29 cells (ATCC, Rockville, MD) were resuspended in the cell medium (McCoy's 5A Modified medium supplemented with 10% FBS, 2 mM glutamine, 0.75% sodium bicarbonate, 100 U/ml penicillin, and 100 U/ml streptomycin), following treatment with Accutase (Innovative Cell Technologies), washed once with cold Dulbecco's phosphate-buffered saline with calcium and magnesium (DPBS++) containing 1% bovine serum albumin (BSA), and then chilled on ice. 100 μL of each toxin/antibody sample was added to separate vials containing 3×10^5 cells. After mixing, samples were incubated on ice in the dark. After 30 min, 1 ml of ice cold DPBS++/1% BSA was added to each sample. To remove unbound toxin, cell suspensions were washed twice with ice cold DPBS++/1% BSA by centrifuging for 5 min at 4°C at 200 × g and removing the supernatant. Washed cells were resuspended in 500 μl cold DPBS++/1% BSA and analyzed by flow cytometry using an LSRII instrument (BD Biosciences) with excitation and emission wavelengths of 488 and 530 nm, respectively. 10,000 events were measured for each sample.

2.3. Rac1 glucosylation assay

Vero cells were seeded at a cell density of 5×10^3 cells/well in a 384-well collagen-coated plate and grown overnight at 37°C in 5% CO_2. TcdA and TcdB (The Native Antigen Company,

Upper Heyford, the UK and tgcBIOMICS, Bingen, Germany) were serially diluted in Vero cell culture medium, and 50 μl was added to each well. For assays determining neutralization effects of actoxumab and bezlotoxumab, TcdA and TcdB were pre-incubated at 90% effective concentrations (EC_{90}) with actoxumab and bezlotoxumab, respectively (various concentrations), for 1 h at RT in Vero cell culture medium, prior to addition of cells as above. Following incubation at 37°C in 5% CO_2 for 3 h, medium containing toxin alone or toxin+antibody was removed by aspiration. Cells were immediately fixed with 50 μl/well fixing solution (4% paraformaldehyde in modified Dulbecco's phosphate-buffered saline (DPBS/modified)) for 1 h at RT. Following fixation, cells were washed four times for 5 min with 50 μl/well permeabilization solution (0.1% Triton-X-100 in DPBS/modified) at RT with gentle shaking. Cells were then blocked with 50 μl/well Odyssey blocking buffer (Li-Cor) overnight at 4°C. After removing blocking buffer, cells were incubated with 25 μl/well mouse anti-Rac1 (BD Biosciences #610651, recognizing non-glucosylated Rac1), or anti-Rac1 clone 23A8 (Millipore #05-389, recognizing total Rac1), diluted at 1:75 and 1:200, respectively, in Odyssey blocking buffer and incubated for 2 h at RT with gentle shaking. Cells were washed four times for 5 min with 50 μl/well wash solution (0.1% tween 20 in DPBS/modified) at RT with gentle shaking. Cells were then incubated with 25 μl/well secondary antibodies (IRDye 800 CW goat anti-mouse and CellTag 700 stain, diluted at 1:800 and 1:1000, respectively, in Odyssey blocking buffer) at RT for 1 h with gentle shaking protected from light. Cells were again washed four times for 5 min with 50 μl/well wash solution at room temperature with gentle shaking. After the final wash, any remaining solution was removed from the wells, and the plates were scanned on the Li-Cor Odyssey classic (Li-Cor) with detection in both 700 and 800 nm channels (A700 and A800). Cell number normalization/well was calculated using the ratio of A800/A700, and remaining percent of non-glucosylated Rac1 was determined using the ratio of normalized A800 of treated cells/normalized A800 of untreated cells multiplied by 100. Analysis was performed with GraphPad Prism (version 6.04) using the 4-parameter nonlinear regression formula.

2.4. Dynamic mass redistribution (Epic) assay

Vero cells were seeded at a cell density of 5×10^3 cells/well in a 384-well fibronectin-coated Epic plate (Corning #5042) and grown overnight at 37°C in 5% CO_2. On the day of assay, medium was aspirated and replaced with 40 μl/well assay buffer (HBSS in 20 mM HEPES) and equilibrated at RT for 1 h. TcdA and TcdB (The Native Antigen Company, Upper Heyford, UK and tgcBIOMICS, Bingen, Germany) were serially diluted in assay buffer and equilibrated at RT for approximately 10 min. For assays determining neutralization effects of actoxumab and bezlotoxumab, TcdA and TcdB were pre-incubated at EC_{90} concentrations for 1 h at RT with actoxumab and bezlotoxumab, respectively (various concentrations). Following pre-incubations, 10 μl/well of the toxins alone or toxin/antibody solutions were added to Vero cells using a Matrix Platemate (Thermo Scientific) and gently mixed. The plate was read every 12 s for 200 min using the Epic BT-157900 (Corning). As a baseline, wells containing assay buffer alone were used. The dynamic mass redistribution (DMR) values were recorded at 180 min at which point the signal had plateaued (not shown). The recorded DMR values (corrected for assay buffer alone) were collected with EpicAnalyzer software and analyzed with GraphPad Prism (version 6.04) using the four-parameter nonlinear regression formula.

http://dx.doi.org/10.5772/68127

2.5. Transepithelial electrical resistance (TER) assay

To initiate the 2-dimensional culture system, 0.5–1 × 10^5 Caco-2 cells (ATCC, Rockville, MD) were seeded into each well of the 24-well insert plates (Falcon #351181 HTS Multiwell Insert System—1.0 um pore size/PET membrane), with 250 µl Caco-2 cell culture medium (EMEM supplemented with 10% FBS, 1× non-essential amino acid, 0.075% sodium bicarbonate, 100 U/ml penicillin, and 100 U/ml streptomycin) in the apical chamber and 800 µl in the basolateral chamber. Caco-2 cells were cultured for at least 14 days at 37°C with 5% CO_2 to ensure full differentiation and confluency, which were confirmed by plateauing of the TER reading at ≥ 600 Ω cm^2. TER was measured using the Epithelial Volt-Ohm Meter Millicell ERS-2 (EMD Millipore, Billerica, MA, USA). To assess the effect of toxins on the cell monolayer, TcdA and TcdB (The Native Antigen Company, Upper Heyford, UK and tgcBIOMICS, Bingen, Germany) were added to the apical chamber. To evaluate the ability of the antibodies to neutralize toxin effects, actoxumab or bezlotoxumab was added to the apical chamber immediately before addition of TcdA or TcdB to the apical chamber. For neutralization studies, 10 ng/ml TcdA was combined with various concentrations (from 0 to 50 µg/ml) of actoxumab, and 100 ng/ml TcdB was combined with various concentrations (from 0 to 100 µg/ml) of bezlotoxumab. TER measurements were obtained immediately before and, at 6, 24, and 48 h, after addition of toxins/antibodies to the apical chamber. TER values were normalized to values obtained in the absence of toxin at each time point to account for minor time-dependent variability.

2.6. Sulforhodamine B assay

To study the effects of *C. difficile* toxins on cytotoxicity and the ability of actoxumab and bezlotoxumab to neutralize those effects, the sulforhodamine B (SRB) assay was employed to measure total cellular protein as a surrogate of cell number [16]. Vero or T-84 (T-84 growth medium—DMEM/F-12K supplemented with 5% FBS, 2 mM L-glutamine, 100 U/ml penicillin, 100 U/ml streptomycin) cells were seeded into 96-well plates at 2000 and 3000 cells/well, respectively, and incubated overnight at 37°C with 5% CO_2. Varying concentrations of purified TcdA and TcdB (tgcBIOMICS, Bingen, Germany) were diluted in the appropriate growth media, incubated at 37°C for 2 h, and added to cells. Following a 24-h incubation at 37°C with 5% CO_2, the medium was aspirated and plates were washed twice with PBS. About 200 µl per well of complete medium was added, and plates were incubated for an additional 48 (Vero cells) or 72 h (T-84 cells). After incubation, the medium was removed, and cells were fixed with 100 µl/well of 10% cold trichloroacetic acid (TCA) for 1 h at 4°C. The TCA was then removed and plates were washed four times with distilled water. After washing, 100 µl/well of 100 µg/ml SRB in 10% acetic acid was added, and plates were incubated for 15 min at room temperature (RT). The plates were then washed four times with 10% acetic acid and air-dried. Addition of 150 µl/well of 10 mM tris was followed by a 10-min incubation at RT with shaking. Absorbance was then measured at 570 nm with a SpectraMax plate reader (Molecular Biosystems). Treated and untreated cells were compared, and 90% lethal concentrations (LC_{90}, that is, concentrations of TcdA or TcdB required to cause a 90% reduction in cell number) were calculated. Antibody-mediated toxin neutralization was measured by incubating serially diluted actoxumab or bezlotoxumab (at concentrations ranging from 1 ng/ml to 192 µg/ml) with purified TcdA or TcdB at LC_{90} for 2 h at 37°C. The toxin/antibody mixtures were then added to Vero or T-84 cells as described above and incubated for 24 h at 37°C with 5% CO_2. The cells were then washed twice with PBS and treated and analyzed as described above.

To assess the cytotoxicity of *C. difficile* toxins derived from bacterial culture supernatants, strain VPI 10463 (ribotype 087) (ATCC) was grown in chopped meat medium (Anaerobe Systems) under anaerobic conditions at 37°C for 72–96 h, and culture supernatants were collected, filtered twice through a 0.22 μm filter, and stored at 4°C. For TcdB immunodepletion, cell culture supernatants were combined and mixed with bezlotoxumab and protein A-agarose beads for 4–6 h at 4°C. After incubation, the beads were removed by centrifugation. Supernatants were then collected, filtered (0.22 μm), and stored at 4°C. Cytotoxicity and antibody-mediated neutralization of the untreated (for determinations on TcdB) or immunodepleted (for determinations on TcdA) supernatants were measured as described above.

3. Results

3.1. Overview of mammalian cell intoxication by TcdA and TcdB

TcdA and TcdB are large, monomeric proteins (300 and 270 kDa, respectively) with similar structures and functions (**Figure 1**) [17, 18]. The functional domains of the toxins are arranged according to the ABCD model [17]: the N-terminal A domain contains the glucosyltransferase enzymatic activity, the B domain is a putative receptor-binding domain composed of a series of long and short repeats known as combined repetitive oligopeptides (CROPs), the cysteine protease (C) domain is responsible for autocatalytic processing, and the D domain is involved in pore formation and toxin translocation. Both toxins bind to receptors on the surface of the epithelial cells that line the wall of the lower intestine (and possibly other cell types). Once bound, they are internalized via receptor-mediated endocytosis [19]. Acidification of the endosome promotes a conformational change that enables translocation of the N-terminal glucosyltransferase domain of the toxin into the cytoplasm. Cellular inositol hexakisphosphate (InsP6) allows cleavage of the toxin by the cysteine protease domain, releasing the glucosyltransferase domain into the cytoplasm where it inactivates Rho-type GTPases through covalent glucosylation (from UDP-glucose) [20]. This in turn causes changes in epithelial cell morphology due to actin depolymerization, loss of tight junction integrity, and eventually, cell death (**Figure 1**) [21]. The assays described in this chapter measure many of the various steps, described above, involved in the intoxication cascade (steps 1–5, as denoted in **Figure 1**).

3.2. Cell surface binding of TcdA and TcdB (step 1 in Figure 1)

Binding of toxins to the cell surface of target cells is the first step in TcdA and TcdB cell entry, leading to the downstream effects of the toxins. We assessed cell surface binding of TcdA and TcdB by Western blotting of cell membranes isolated from Vero cells incubated

http://dx.doi.org/10.5772/68127

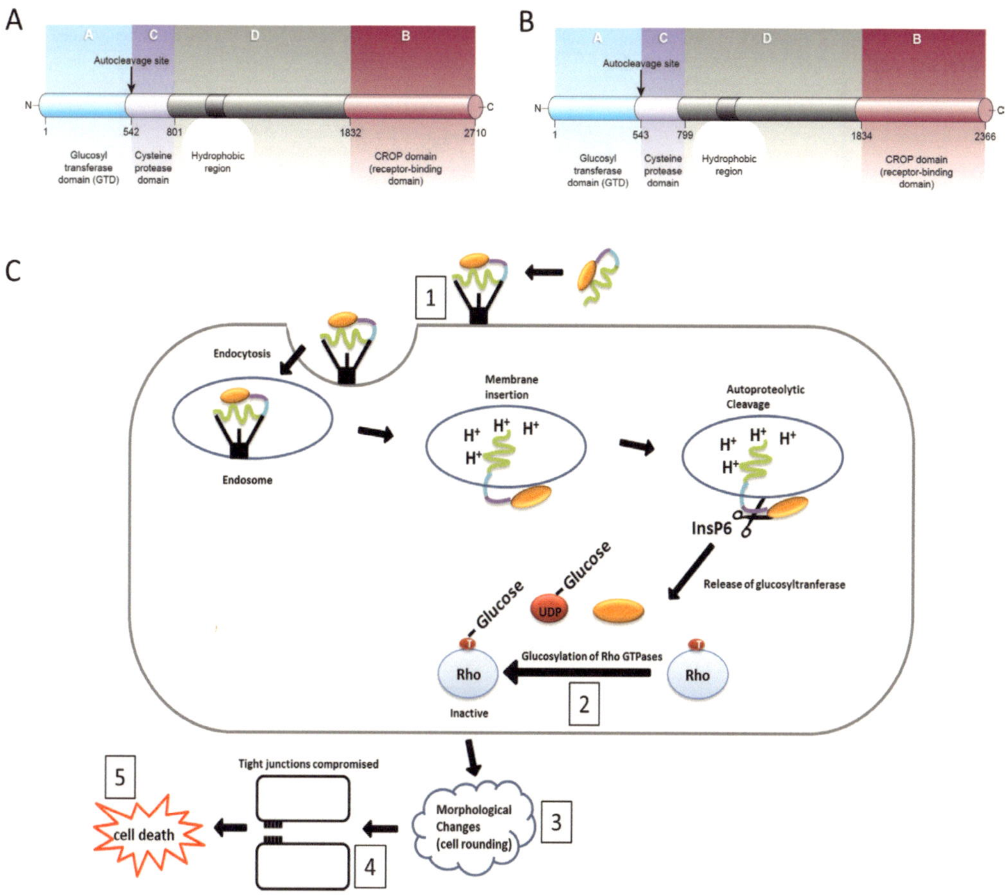

Figure 1. *Clostridium difficile* toxin structure and mechanism of action. (A) Domain organization of TcdA. (B) Domain organization of TcdB. (C) Mechanism of intoxication of mammalian cells by TcdA and TcdB. Toxins A and B bind to receptors on the surface of target cells (1) and are endocytosed. Endosomal toxins are acidified causing exposure of hydrophobic regions of the protein that allow their insertion into the membrane, forming pore(s). The N-terminal catalytic domain is then translocated from the endosomal compartment into the cytoplasm, where the glucosyltransferase domain is released by inositol hexakisphosphate (InsP6)-dependent auto-cleavage. The toxins then glucosylate Rho-type GTPases (2) from UDP-glucose, causing actin depolymerization, changes in cell morphology (3), disruption of tight junctions (4), and cell death (5). Cellular events numbered 1–5 correspond to the steps assessed by the various assays described in this chapter. Figure adapted from Jank and Aktories [17].

with TcdA or TcdB at 4°C. As shown in **Figure 2**, membrane fractions isolated from cells incubated with TcdA (see **Figure 2A**, top panel) or TcdB (**Figure 2B**, top panel) contain toxins, indicating cell surface binding of the toxins. Actoxumab and bezlotoxumab bind to and neutralize purified TcdA and TcdB, respectively, from a variety of *C. difficile* strains [15]. Pre-incubation of TcdA with actoxumab but not bezlotoxumab efficiently blocked binding of TcdA to cells (**Figure 2A**), while pre-incubation of TcdB with bezlotoxumab but not actoxumab efficiently blocked binding of TcdB to cells (**Figure 2B**), confirming the specificity of toxins binding to cells.

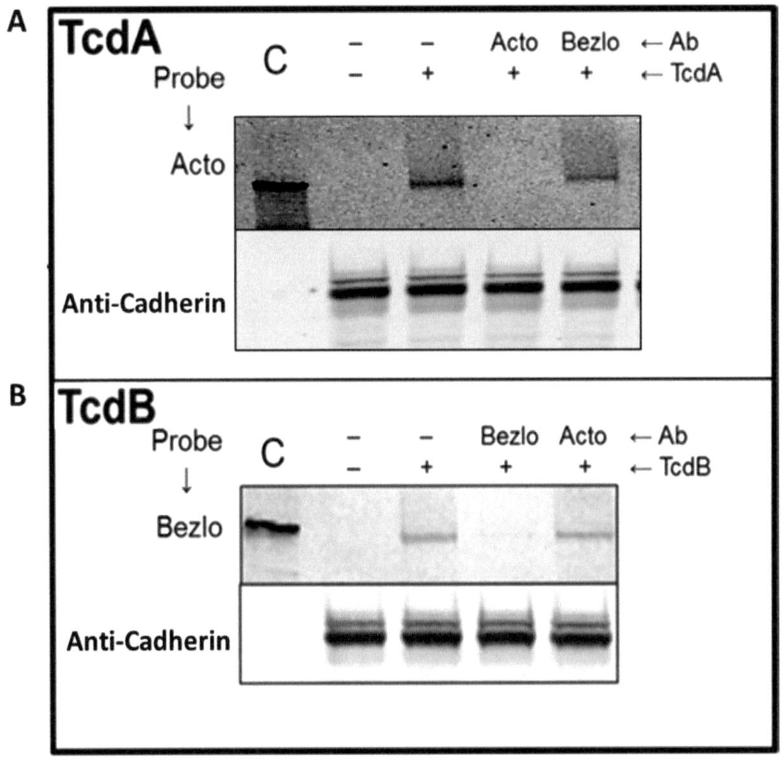

Figure 2. Cell surface binding of TcdA and TcdB as measured by Western blot. Western blots of cell membranes isolated from Vero cells following incubation with (A) TcdA or (B) TcdB, in the presence of vehicle, actoxumab, or bezlotoxumab (200 μg/ml), as indicated. The top blots in each panel show TcdA and TcdB, while the bottom blots show cadherin, used as a loading control.

Binding of TcdA to cells and the prevention, thereof, by actoxumab were also assessed by flow cytometry (**Figure 3**). Incubation of HT29 cells with increasing levels of fluorescently labeled TcdA (TcdA-Atto488) led to an elevated mean fluorescence intensity (MFI), indicating binding of TcdA to the cell surface in a concentration-dependent manner. In the presence of actoxumab, however, the MFI for each toxin concentration was reduced to background levels showing that actoxumab blocked binding of TcdA to the cell surface. No significant changes in MFI were measured in the presence of bezlotoxumab, indicating that the effect of actoxumab is specific (data not shown).

3.3. Glucosylation of Rac1 by TcdA and TcdB (step 2 in Figure 1)

Inactivation of Rho-type GTPases is a key step in the intoxication of host cells, leading to the downstream cytopathic and cytotoxic effects of the *C. difficile* toxins. Historically, the glucosylation of Rho GTPases was assessed by polyacrylamide gel-based assays that use either radioactively labeled glucose or antibodies to detect the glucosylated and non-glucosylated protein on a gel [22]. These assays are laborious, low throughput, qualitative, and do not detect glucosylation directly in the cell. A novel assay was therefore developed to measure

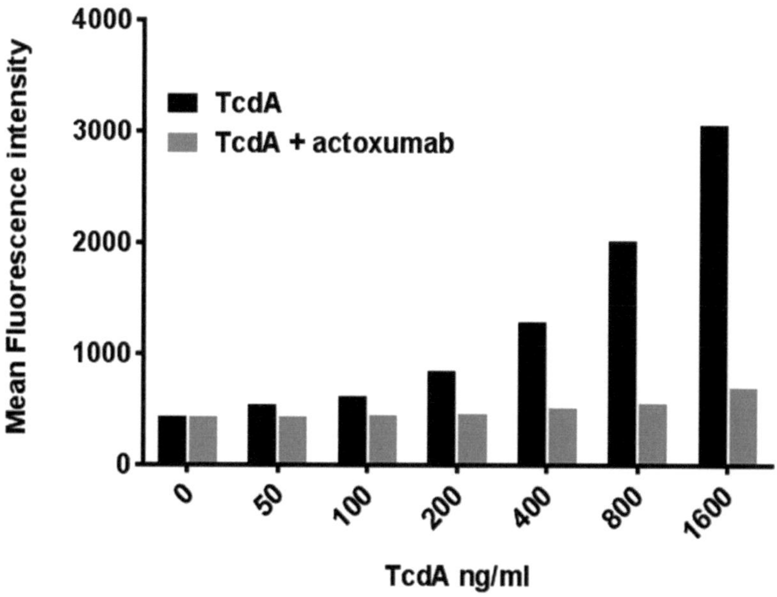

Figure 3. Cell surface binding of TcdA as measured by flow cytometry. A representative experiment showing flow cytometry analysis of HT29 cells pre-incubated with a titration of TcdA-Atto488 in the presence or absence of actoxumab. Following incubation, mean fluorescence intensity (MFI) was measured with excitation and emission wavelengths of 488 and 530 nm, respectively.

TcdA- and TcdB-mediated Rac1 glucosylation in a high throughput and quantitative 384-well in-cell Western assay, using antibodies that detect non-glucosylated and total Rac1. A dose-dependent decrease in non-glucosylated Rac1 was observed in the presence of TcdA and TcdB from various ribotypes (027, 078, and the control 087 (strain VPI 10463)) (**Figure 4A** and **B**), while total Rac1 was minimally affected (not shown). Vero cells were found to be more sensitive to TcdB than TcdA, consistent with previous observation by Torres et al. [23]. In addition, differences in sensitivity of Vero cells to toxins of the different *C. difficile* ribotypes were noted. For instance, Vero cells were found to be more sensitive to TcdA of ribotype 087 (VPI 10463) than of ribotypes 027 and 078, while TcdB showed the opposite effect, with cells being more sensitive to TcdB of ribotypes 027 and 078 compared to ribotype 087.

Actoxumab and bezlotoxumab neutralized the effects of TcdA and TcdB (at EC_{90} concentrations), respectively (**Figure 4C** and **D**). Notably, the potency of actoxumab and bezlotoxumab on their respective toxins was lower for toxins of ribotype 027 and 078 compared to ribotype 087. This is consistent with the lower affinities of the antibodies against toxins of these ribotypes, as previously described by Hernandez et al. [15].

3.4. Changes in cell morphology induced by TcdA and TcdB (step 3 in Figure 1)

The cytopathic effects of TcdA and TcdB on gut epithelium are visualized as profound morphological changes, typically cell rounding, due to the glucosylation and inactivation of Rho-type GTPases and subsequent disruption of actin polymerization. Historically, these cytopathic

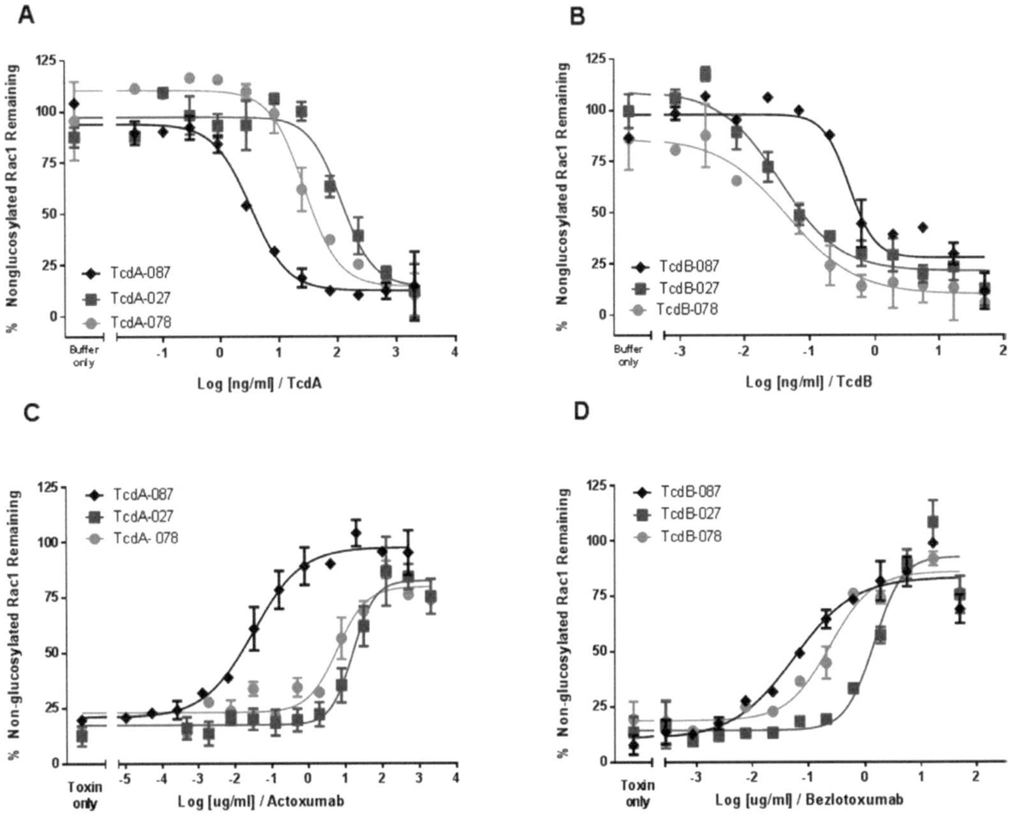

Figure 4. TcdA- and TcdB-mediated Rac1 glucosylation and neutralization thereof by actoxumab and bezlotoxumab. Effect of TcdA (A) and TcdB (B) of ribotypes 027, 078, and 087 on glucosylation of Rac1. Neutralization of TcdA-mediated Rac1 glucosylation by actoxumab (C) and of TcdB-mediated Rac1 glucosylation by bezlotoxumab (D).

effects have been assessed qualitatively through visual determination of cell rounding [23]. Improved phenotypic assays used to investigate changes in cell morphology involve the quantification of length-to-width ratios of fluorescently labeled cells [23, 24]. This latter technique is quantitative and has an improved throughput, although it requires consistent staining and substantial data analysis. To better understand and quantify toxin-induced morphological changes in unlabeled cells, an assay was developed to examine dynamic mass distribution (DMR) in Vero cells using the Epic instrument. In this assay, plates containing optical sensors are used to capture translocation of cellular mass of unlabeled cells in response to ligand binding, allowing changes in cell shape to be quantified. The concentration-dependent effects of TcdA and TcdB on mass redistribution were determined at 180 min (at which time the effects have plateaued, not shown) (**Figure 5A**). As with the Rac1 glucosylation assay, Vero cells are much more sensitive to TcdB than TcdA in the DMR assay. The neutralizing effects of actoxumab and bezlotoxumab on toxin-induced morphological changes were assessed at EC_{90} concentrations of TcdA and TcdB, respectively. Actoxumab and bezlotoxumab fully neutralized the effects of TcdA and TcdB, respectively, on DMR (**Figure 5B**).

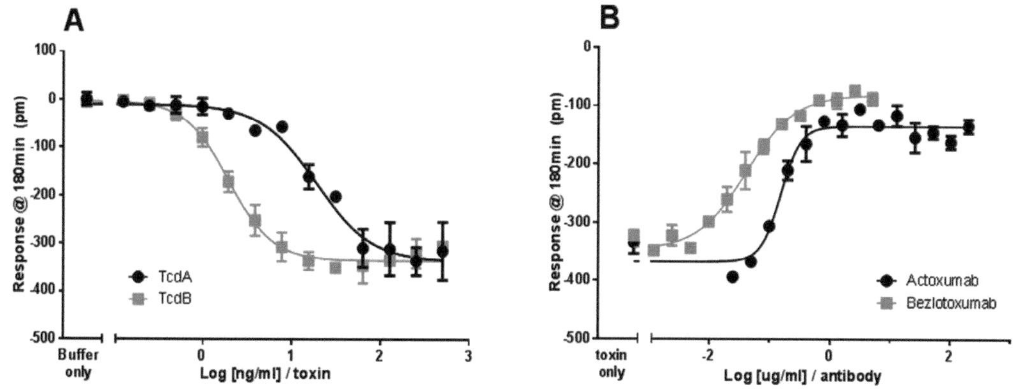

Figure 5. Effects of TcdA and TcdB on dynamic mass redistribution and neutralization by actoxumab and bezlotoxumab. (A) Concentration-dependent effects of TcdA and TcdB on DMR. (B) Neutralizing effects of actoxumab and bezlotoxumab on toxin-induced effects on DMR.

3.5. Toxin-induced disruption of epithelial tight junctions (step 4 in Figure 1)

To gain an understanding of the effect of *C. difficile* toxins on the integrity of the gut wall epithelium, a two-dimensional cell culture system was utilized wherein a single monolayer of colonic epithelial cells (Caco-2) is grown on a permeable membrane, separating distinct apical and basolateral compartments [25–28]. The system simulates the polarized nature of the intact intestinal mucosal epithelium, which separates the gut lumen (apical side) from the subepithelial/systemic space (basolateral side). The integrity of the epithelial layer is monitored by measuring the transepithelial electrical resistance (TER), with a decrease in TER suggesting that the integrity of the epithelial monolayer has been compromised [26]. In this system, TcdA and TcdB added to the apical side of the cell monolayer (mimicking the presence of toxin on the lumenal side of the gut) caused significant time- and concentration-dependent decreases in TER (**Figure 6A** and **B**). Neutralization of the toxin-induced effects by actoxumab and bezlotoxumab was assessed at EC_{90} concentrations of TcdA and TcdB, respectively. Both antibodies dose-dependently neutralized the effects of their respective toxins (**Figure 6C** and **D**).

3.6. Toxin-induced cytotoxicity (step 5 in Figure 1)

The traditional way of assessing the cytoxic effects of *C. difficile* on host cells involves measuring cellular ATP levels of intoxicated cells. This method is plagued with low signal to noise ratios and variability due to substantial ATP levels remaining in cells that are not yet dead and still undergoing morphological changes due to intoxication [13]. Additionally, normal metabolism-related fluctuations in ATP levels that are unrelated to cell viability can further affect the assay readout. We developed a more robust colorimetric assay that measures cellular protein content as a surrogate of cell growth and survival [14]. The sulforhodamine B (SRB) assay was used to determine the cytotoxic effects of purified *C. difficile* toxins of the reference strain VPI 10463 (ribotype 087) and from strains of ribotypes from the so-called hyper-virulent ribotypes 027 and 078. All toxins tested caused a robust concentration-dependent decrease in cell viability

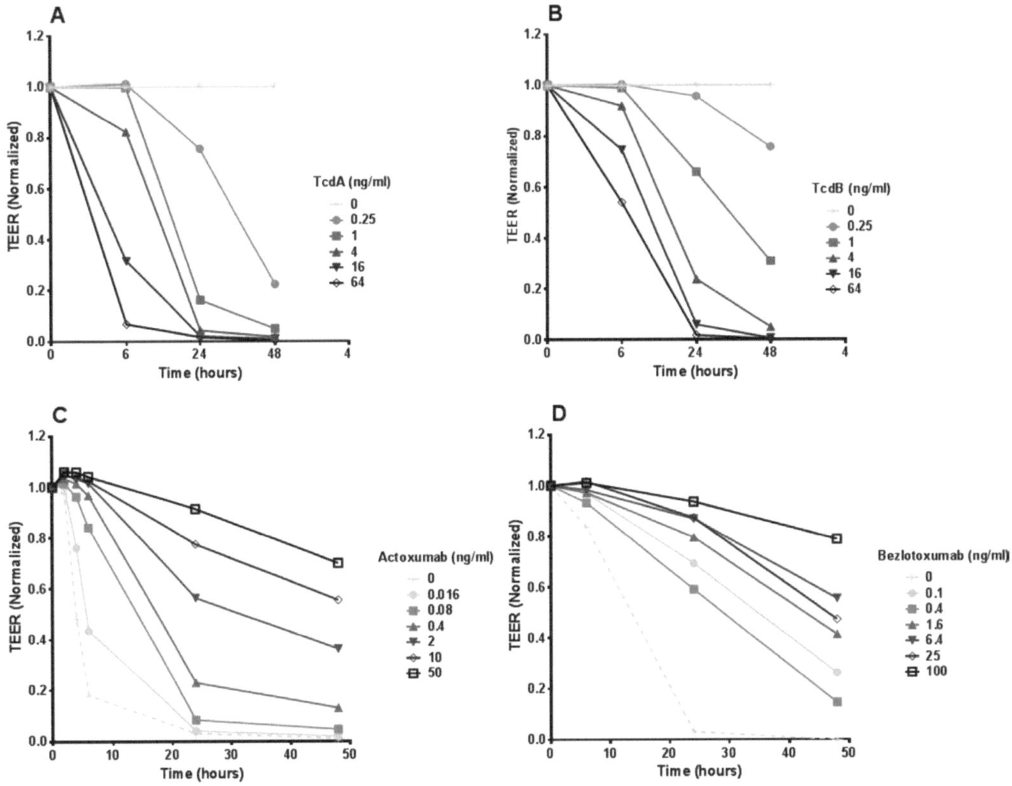

Figure 6. Effects of TcdA and TcdB on integrity of Caco-2 cell monolayers and neutralization by actoxumab and bezlotoxumab. Time- and concentration-dependent effects on TER of TcdA (A) or TcdB (B) added to the apical side of Caco-2 monolayers. Time- and dose-dependent neutralization of TcdA by actoxumab (C) and of TcdB by bezlotoxumab (D), added to the apical side.

(**Figure 7A** and **B**). As with other assays described herein and as previously observed by Torres et al. [23], Vero cells are significantly more sensitive to TcdB than to TcdA. The ability of actoxumab and bezlotoxumab to neutralize TcdA and TcdB, respectively, was assessed at toxin concentrations that are associated with a 90% decrease in cell viability (LC_{90}). Both antibodies fully neutralized the effects of their respective toxins from all ribotypes tested (**Figure 7C** and **D**). However, the neutralization potencies of both antibodies for toxins of ribotypes 027 and 078 were significantly lower than toxins of ribotype 087, similar to data obtained in the Rac1 glucosylation assay above (Section 2.3) and consistent with previous data in the SRB assay [15].

The robust nature of the SRB assay also allows for the study of the cytotoxic effects of unpurified *C. difficile* toxins directly from culture supernatants for clinical strains for which purified toxins are not available. For these studies, Vero cells were treated with serially diluted culture supernatants of the reference strain VPI 10463, containing both toxins (not shown), in the absence or presence of actoxumab, bezlotoxumab, or the combination of both antibodies. In the absence of antibodies, there was a concentration-dependent decrease in cell viability, presumably due to the presence of toxin in the supernatant. Addition of actoxumab had no effect on the cytotoxicity of supernatant, while addition of 10 μg/ml bezlotoxumab either

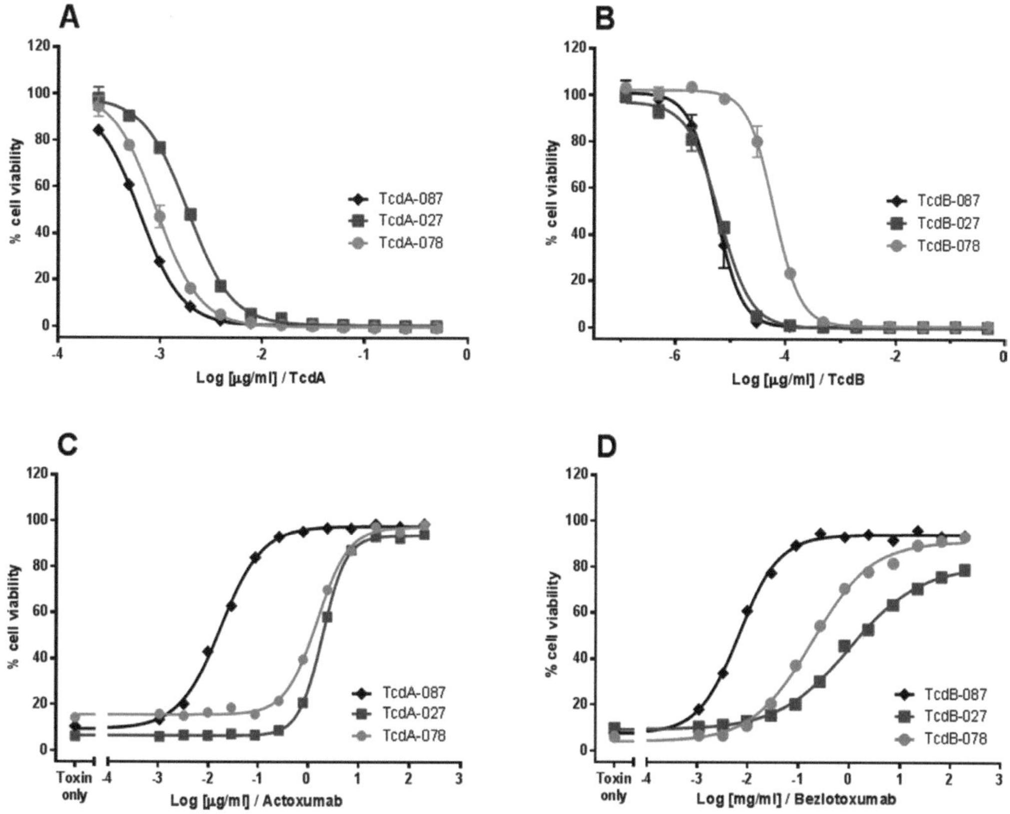

Figure 7. Purified TcdA- and TcdB-mediated effects on cell viability and neutralization by actoxumab and bezlotoxumab. Reduction in Vero cell viability induced by TcdA (A) and TcdB (B) using purified toxins from ribotypes 087, 027, and 078. Neutralization of TcdA by actoxumab (C) and of TcdB by bezlotoxumab (D). Figure reproduced from Hernandez et al. [15] (Copyright © American Society for Microbiology [Antimicrob Agents Chemother. 59, 2015, 1052–1060. DOI:10.1128/AAC.04433-14]).

by itself or in combination with 10 µg/ml actoxumab significantly shifted the concentration-response curve to the right, indicating that most of the cytotoxic activity in the supernatant is due to TcdB (**Figure 8A**). This is not surprising as Vero cells are more sensitive to TcdB than to TcdA. To assess the cytotoxic activity associated with TcdA, TcdB was first removed from the supernatant using an immunodepletion approach (see Section 2). In this case, 10 µg/ml actoxumab, alone or in combination with 10 µg/ml bezlotoxumab, shifted the response curve to the right, whereas bezlotoxumab showed minimal effect, confirming that the cytotoxic activity in immunodepleted supernatants is associated mainly with TcdA (**Figure 8B**). To confirm this finding, full concentration-response curves of actoxumab and bezlotoxumab were generated against dilutions of intact or immunodepleted supernatants associated with ~90% reduction in cell viability (EC_{90}); actoxumab neutralized the cytotoxic activity of immunodepleted supernatants, whereas bezlotoxumab neutralized the cytotoxic activity of intact supernatants, and no cross-neutralization was observed (**Figure 8C** and **D**). This approach has been used successfully to assess the activities of actoxumab and bezlotoxumab on TcdA and TcdB of dozens of clinical isolates of *C. difficile*, covering 18 distinct ribotypes (seven toxinotypes) [15].

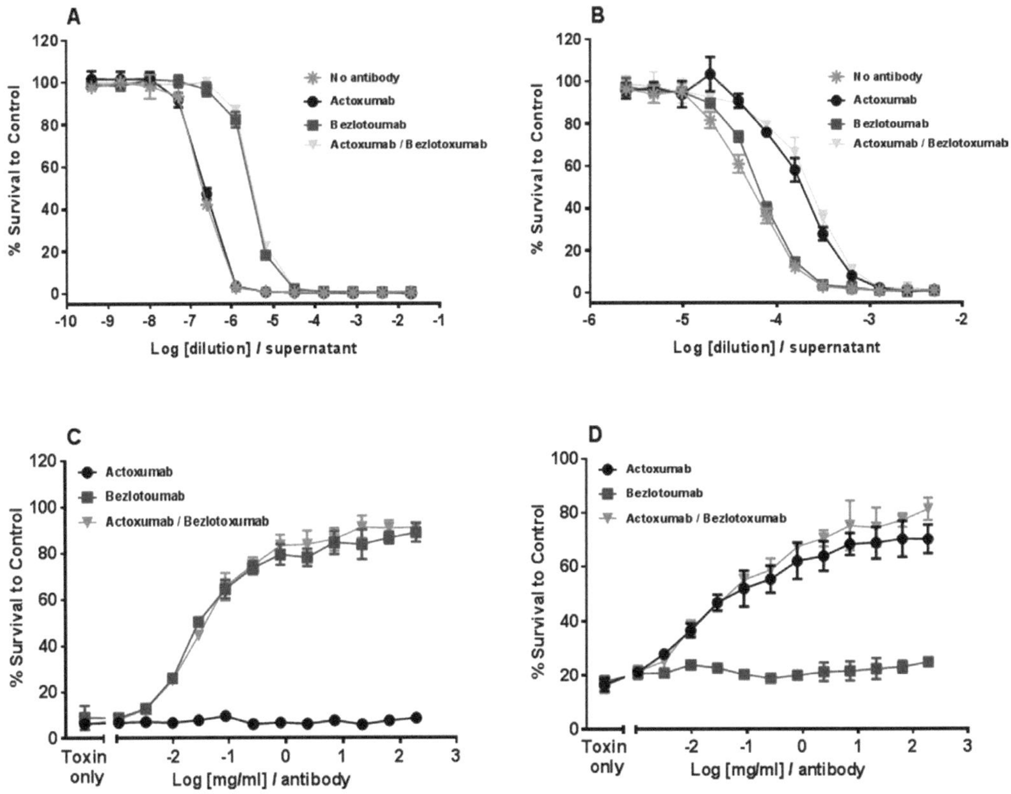

Figure 8. Unpurified TcdA- and TcdB-mediated effects on Vero cell viability and neutralization by actoxumab and bezlotoxumab. Cytotoxic effects of serially diluted intact (A) or immunodepleted (B) supernatants in the presence or absence of actoxumab, bezlotoxumab, or a combination of the two antibodies. Neutralization of cytotoxic activity by bezlotoxumab, but not actoxumab, in intact supernatant at EC_{90} dilution (C) and by actoxumab but not bezlotoxumab in immunodepleted supernatant at EC_{90} dilution (D).

4. Conclusions

In this chapter, we have described novel cell-based assays for analyzing multiple distinct steps in the intoxication cascade associated with TcdA and TcdB. Unlike historical assays that measure toxin effects qualitatively, such as the visual assessment of cell rounding, or are variable and often unreliable, such as quantitation of ATP levels to estimate cell death, the assays presented here can quantitatively and robustly assess the effects of toxins in mammalian cells. We show how the initial event of toxin binding to host cells can be assessed using cell surface binding assays with labeled or unlabeled toxins in flow cytometry and Western blot formats, respectively. The more proximal events that follow internalization of the toxins, namely Rac1 glucosylation and cell rounding, can be studied with novel quantitative assays by in-cell Western and dynamic mass redistribution assays, respectively. Finally, we show how the TER and SRB assays can be utilized to assess the final stages of intoxication, tight junction disruption, and cell death, respectively. We also show how the SRB assay can be used to accurately measure the activities of TcdA and TcdB from unpurified toxins in culture supernatants of

C. difficile strains for which purified toxins are not available. The assays described were validated with the antitoxin antibodies actoxumab (anti-TcdA) and bezlotoxumab (anti-TcdB) to demonstrate their utility in evaluating pharmacological blockade of toxins. These assays may be useful in future studies aimed at better understanding of *C. difficile* toxin function, as well as in characterizing toxin inhibitors as tools or as potential therapeutics.

Author details

Mary Ann Cox, Lorraine D. Hernandez, Pulkit Gupta, Zuo Zhang, Fred Racine and Alex G. Therien* †

*Address all correspondence to: atherien@inceptionsci.com

Merck Research Laboratories, Merck & Co., Inc., Kenilworth, NJ, USA

† Current affiliation: Inception Sciences Canada, Montreal, Quebec, Canada

References

[1] Bassetti M, Villa G, Pecori D, Arzese A, Wilcox M. Epidemiology, diagnosis and treatment of *Clostridium difficile* infection. Expert Rev Anti Infect Ther. 2012;10:1405-23. DOI: 10.1586/eri.12.135

[2] Rupnik M, Wilcox MH, Gerding DN. *Clostridium difficile* infection: new developments in epidemiology and pathogenesis. Nat Rev Microbiol. 2009;7:526-36. DOI:10.1038/nrmicro 2164

[3] Carter GP, Rood JI, Lyras D. The role of toxin A and toxin B in the virulence of *Clostridium difficile*. Trends Microbiol. 2012;20:21-9. DOI: 10.1016/j.tim.2011.11.003

[4] Dingle T, Wee S, Mulvey GL, Greco A, Kitova EN, Sun J, et al. Functional properties of the carboxy-terminal host cell-binding domains of the two toxins, TcdA and TcdB, expressed by *Clostridium difficile*. Glycobiology. 2008;18:698-706. DOI:10.1093/glycob/cwn048

[5] Pruitt, RN, Lacy DB. Toward a structural understanding of *Clostridium difficile* toxins A and B. Front Cell Infect Microbiol. 2012;2:28. DOI:10.3389/fcimb.2012.00028

[6] Papatheodorou P, Zamboglou C, Genisyuerek S, Guttenberg G, Aktories K. Clostridial glucosylating toxins enter cells via clathrin-mediated endocytosis. PLoS One. 2010;5:e10673. DOI:10.1371/journal.pone.0010673

[7] Pfeifer G, Schirmer J, Leemhuis J, Busch C, Meyer DK, Aktories K, et al. Cellular uptake of *Clostridium difficile* toxin B. Translocation of the N-terminal catalytic domain into the cytosol of eukaryotic cells. J Biol Chem. 2003;278:44535-41. DOI:10.1074/jbc.M307540200

[8] Shen A. *Clostridium difficile* toxins: mediators of inflammation. J Innate Immun. 2012; 4:149-58. DOI:10.1159/000332946

[9] Cohen SH, Gerding DN, Johnson S, Kelly CP, Loo VG, McDonald LC, Pepin J, Wilcox MH. Society for Healthcare Epidemiology of America, Infectious Diseases Society of America. Clinical practice guidelines for *Clostridium difficile* infection in adults: 2010 update by the society for healthcare epidemiology of America (SHEA) and the infectious diseases society of America (IDSA). Infect Control Hosp Epidemiol. 2010;31:431-55.

[10] Venugopal AA, Johnson, S. Current state of *Clostridium difficile* treatment options. Clin Infect Dis. 2012;55:S71–S76. DOI: 10.1093/cid/cis355

[11] Stamper P, Alcabasa R, Aird D, Babiker W, Wehrlin J, Ikpeama I, Carroll K. Comparison of a commercial real-time PCR assay for *tcdB* detection to a cell culture cytotoxicity assay and toxigenic culture for direct detection of toxin-producing *Clostridium difficile* in clinical samples. J Clin Microbiol. 2009;47:373-78. DOI: 10.1128/JCM.01613-08

[12] Babcock, GJ, Broering TJ, Hernandez HJ, Mandell RB, Donahue K, Boatright N, et al. Human monoclonal antibodies directed against toxins A and B prevent *Clostridium difficile* induced mortality in hamsters. Infect Immun. 2006;74:6339-47. DOI:10.1128/IAI.00982-06

[13] Xie J, Zorman J, Indrawati L, Horton M, Soring K, Antonello J, Zhang Y, Secore S, Miezeiewski M, Wang S, Kanavage A, Skinner J, Rogers I, Bodmer J, Heinrichs J. Development and optimization of a novel assay to measure neutralizing antibodies against *Clostridium difficile* toxins. Clin Vaccine Immunol. 2013;20:517-25. DOI:10.1128/CVI.00549-12

[14] Orth P, Xiao L, Hernandez LD, Reichert P, Sheth PR, Beaumont M, Yang X, Murgolo N, Ermakov G, DiNunzio E, Racine F, Karczewski J, Secore S, Ingram RN, Mayhood T, Strickland C,Therien AG. Mechanism of action and epitopes of *Clostridium difficile* toxin B-neutralizing antibody bezlotoxumab revealed by X-ray crystallography. J Biol Chem. 2014. 289:18008-21. DOI: 10.1074/jbc.M114.560748

[15] Hernandez LD, Racine F, Xiao L, DiNunzio E, Hairston N, Sheth PR, Murgolo NJ, Therien AG. Broad coverage of genetically diverse strains of *Clostridium difficile* by actoxumab and bezlotoxumab predicted by *in vitro* neutralization and epitope modeling. Antimicrob Agents Chemother. 2015;59:1052-60. DOI:10.1128/AAC.04433-14

[16] Skehan P, Storeng R, Scudiero D, Monks A, McMahon J, Vistica D, et al. New colorimetric cytotoxicity assay for anticancer-drug screening. J Natl Cancer Inst. 1990;82:1107-12. DOI:10.1093/jnci/82.13.1107

[17] Jank T, Aktories K. Structure and mode of action of clostridial glucosylating toxins: the ABCD model. Trends Microbiol. 2008;16:222-9. DOI:10.1016/j.tim.2008.01.011

[18] Davies AH, Roberts AK, Shone CC, Acharya KR. Super toxins from a super bug: structure and function of *Clostridium difficile* toxins. Biochem J. 2011,436:517-26. DOI: 10.1042/BJ20110106

[19] Jank T, Giesemann T, Aktories K. Rho-glucosylating *Clostridium difficile* toxins A and B: new insights into structure and function. Glycobiology. 2007;17:15R–22R. DOI:10.1093/glycob/cwm004

[20] Aktories K. Bacterial protein toxins that modify host regulatory GTPases Nat Rev Microbiol. 2011;9:487-95. DOI:10.1038/nrmicro2592

[21] May M, Wang T, Müller M, Genth H. Difference in F-Actin depolymerization induced by toxin B from the *Clostridium difficile* strain VPI 10463 and toxin B from the variant *Clostridium difficile* Serotype F Strain 1470. Toxins 2013;5:106-19. DOI:10.3390/toxins5010106

[22] Quesada-Gómez C, López-Ureña D, Chumbler N, Kroh HK, Castro-Peña C, Rodríguez C, Orozco-Aguilar J, González-Camacho S, Rucavado A, Guzmán-Verri C, Lawley TD, Lacy DB, Chaves-Olarte E. Analysis of TcdB proteins within the hypervirulent clade 2 reveals an impact of RhoA glucosylation on *Clostridium difficile* proinflammatory activities. Infect Immun. 2016;84:856-65. DOI: 10.1128/IAI.01291-15

[23] Torres J, Camorlinga-Ponce M, Munoz O. Sensitivity in culture of epithelial cells from rhesus monkey kidney and human colon carcinoma to toxins A and B from *Clostridium difficile*. Toxicon. 1992;30:419-26. DOI: 10.1016/0041-0101(92)90538-G

[24] Tam J, Beilhartz G, Auger A, Gupta P, Therien A, Melnyk R. Small molecule inhibitors of *Clostridium difficile* toxin B-induced cellular damage. Chem Biol. 2015;22:175-85. DOI: 10.1016/j.chembiol.2014.12.010

[25] Du T, Alfa MJ. Translocation of *Clostridium difficile* toxin B across polarized Caco-2 cell monolayers is enhanced by toxin A. Can J Infect Dis. 2004;15:83-88.

[26] Nusrat A, von Eichel-Streiber C, Turner JR, Verkade P, Madara JL, Parkos CA. *Clostridium difficile* toxins disrupt epithelial barrier function by altering membrane microdomain localization of tight junction proteins. Infect Immun. 2001;69:1329-36. DOI:10.1128/IAI.69.3.1329-1336.2001

[27] Sambuy Y, De Angelis I, Ranaldi G, Scarino ML, Stammati A, Zucco F. The Caco-2 cell line as a model of the intestinal barrier: influence of cell and culture-related factors on Caco-2 cell functional characteristics. Cell Biol Toxicol. 2005;21:1-26. DOI:10.1007/s10565-005-0085-6

[28] Sutton PA, Li S, Webb J, Solomon K, Brazier J, Mahida YR. Essential role of toxin A in C. difficile 027 and reference strain supernatant-mediated disruption of Caco-2 intestinal epithelial barrier function. Clin Exp Immunol. 2008;153:439-47. DOI: 10.1111/j.1365-2249.2008.03690.x.

Chapter 6

Clostridium difficile in the ICU

William C. Sherman, Chris Lewis, Jong O. Lee and
David N. Herndon

Additional information is available at the end of the chapter

http://dx.doi.org/10.5772/intechopen.69212

Abstract

Clostridium difficile has become an increasingly common infectious agent in the healthcare setting. It is generally associated with antibiotic use and causes diarrhea as well as other complications such as pseudomembranous colitis (PMC) and toxic megacolon. This organism poses a serious threat to patients in the intensive care unit (ICU) as it increases hospital length of stay, morbidity, and mortality. Recurrence rates are typically higher in the ICU population as those patients usually have immunocompromised systems, more exposure to antibiotics and proton pump inhibitors, loss of normal nutritional balance, and alterations in their colonic flora. Emergence of more virulent and pathogenic strains has made combating the infection even more difficult. Newer therapies, chemotherapeutic agents, and vaccinations are on the horizon. However, the most effective treatments to date are ceasing the inciting agent, reduction in the use of proton pump inhibitors, and prevention of the disease. In this chapter, we will explore the risk factors, diagnosis, treatment, and prevention of *C. difficile* infections (CDI) in the ICU.

Keywords: *Clostridium difficile*, intensive care unit, pseudomembranous colitis, toxic megacolon, NAP1

1. Introduction

Clostridium difficile is a gram-positive, spore forming anaerobic bacillus that can survive on environmental surfaces for years in the spore (dormant) stage. First cultured in 1935 by Hall and O'Toole, *C. difficile* was a relatively unknown organism until 1978 [1]. It was initially thought to be a mostly harmless colonizer of the human intestinal tract. In 1893, a young woman died after gastric surgery from a "diphtheric colitis" as described by John Finney and Sir William Osler [2]. In 1978, Dr. John G. Bartlett determined that *C. difficile* was associated with the ailment that had killed the young woman 85 years prior and was now termed pseudomembranous colitis (PMC) [3].

C. difficile is currently the most common cause of antibiotic-associated pseudomembranous colitis in the healthcare setting and caused 20–30% of those with uncomplicated antibiotic-associated diarrhea [4]. According to the Centers for Disease Control, the number of cases of *C. difficile* infections (CDIs) in patients discharged from acute-care facilities doubled from 149,000 to 300,000 between 2001 and 2005 and based on recent trends has reached nearly 500,000 cases per year [5, 6]. There are occasionally other causes of antibiotic associated colitis due to organisms such as *Staphylococcus aureus, Klebsiella oxytoca*, enterotoxin-producing strains of *Clostridium perferingens*, or *Salmonella* [7]. Treatment duration for most microbial infections is usually around 14 days but prolonged exposure to broad-spectrum antibiotics has been associated with increased rates of both initial *C. difficile* infection and recurrence of *C. difficile* infection [8, 9].

The damage caused by *C. difficile* is due to the ability of the microbe to attach to the mucosa of the colon and release of exotoxins into the mucosa. The toxins may cause diarrhea, dilation of the colon (toxic megacolon), (**Figure 1**) sepsis, and death. Transmission is person to person via the fecal-oral route with ingestion of spores that germinate into vegetative bacteria within the small intestine. *C. difficile* produces two toxins—toxins A and B. These are large proteins (308 and 270 kDa, respectively) that cause severe inflammation and necrosis of the mucosal tissue by inactivating Rho, Rac, and Cdc42 targets within the epithelial cells through irreversible glycosylation [10, 11]. Toxin B is thought to be a gene duplication event of toxin A but is 10 times more cytotoxic than toxin A [12, 13].

The bacteria are normally found in up to 25% of hospitalized adults and up to 70% of the hospitalized pediatric population [14]. It does not cause disease until the normal flora is disrupted and *C. difficile* is allowed to proliferate. *C. difficile* infection has a very high economic cost

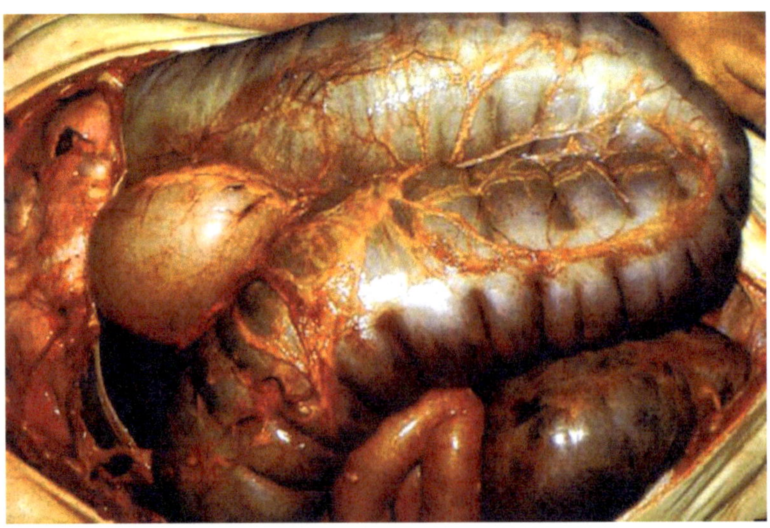

Figure 1. Toxic megacolon related to *Clostridium difficile* infection. Credit: University of Pittsburgh Department of Pathology.

associated with it in the United States and Europe due to high reinfection rates of approximately 30% and risk of relapse of 60% producing over 900,000 cases and an estimated $1.1–$3.2 billion per annum burden [15, 16].

Antibiotic therapy that disrupts the normal flora are usually to blame but proton pump inhibitors and other gastric acid suppression medications are increasingly associated with increases in *C. difficile* overgrowth [17]. Although the cephalosporin class, clindamycin, and the fluoroquinolones are all thought to place a patient at a higher risk of infection, all antibiotics, including oral vancomycin and metronidazole, can induce pseudomembranous colitis due to their ability to eliminate most normal intestinal flora in combination with the increased resistance patterns of more virulent strains of *C. difficile* [3, 14, 18, 19] The NAP1 strain is particularly important as it is associated with fluoroquinolone use and has risen in incidence in Canada, Europe, and the United States with increased virulence, toxin production, mortality, treatment failures, and relapse [20, 21].

The incidence and virulence of this pathogen has been steadily increasing over the last several decades contributing to higher morbidity and mortality. The increasingly older patient population with its higher acuity of medical issues and immunosenescence, the increased use of proton pump inhibitors, and the continued use of antibiotics has all allowed *C. difficile* to leave a greater impact in healthcare settings. In this chapter, we will explore the risk factors, diagnosis, treatment, and prevention of *C. difficile* infections in the intensive care unit (ICU).

2. A historical perspective on *Clostridium difficile*

Pseudomembranous colitis became a common complication of antibiotic use in the 1950s at the beginning of the antibiotic era and was found often in postoperative patients with an incidence of 14–27% [22, 23]. *S. aureus* was the suspected pathogen and standard treatment became oral vancomycin [24].

Tedesco et al. described "clindamycin colitis" in 1974 utilizing culture and endoscopy to diagnose pseudomembranous colitis associated with antibiotic use after 21% of patients given clindamycin developed diarrhea and 10% developed pseudomembranous colitis [25]. Incidentally, *S. aureus* did not grow from stool cultures from any of the patients. This study, more than prior publications, crystallized the connection between antibiotic use and development of pseudomembranous colitis. Green, while studying penicillin-induced death in guinea pigs in 1974 described stool cytopathic changes that he attributed to the activity of a latent virus. In retrospect, this appears to be the first identification of the effects of *C. difficile* cytotoxin [26]. Between 1977 and 1979, using hamster models, multiple teams of researchers identified *C. difficile* as the causative agent of pseudomembranous colitis, including detecting toxin B produced by *C. difficile* [27–30]. "Clindamycin colitis" became known as "antibiotic-induced colitis" and most of the studies done in the 1980s demonstrated that cephalosporins were the most frequently implicated agents followed secondly by broad-spectrum penicillins, including amoxicillin [30–33].

Although there are many causes of pseudomembranous colitis, the majority of cases since the late 1970s have been caused by *C. difficile* infection. Pseudomembranous colitis is limited to the proximal colon in 20–30% of cases and may therefore be missed by sigmoidoscopy, providing more credence to performing a complete colonoscopy to identify anatomic lesions [25, 34]. With the current availability of *C. difficile* toxin assays, colonoscopy is rarely necessary. The first test used to diagnose *C. difficile* involved neutralization of the cytotoxin by *C. sordellii* antitoxin. This remains the most sensitive and specific diagnostic test, but is expensive and requires 24–48 hours for results [35] that has led to the development of latex particle agglutination [36–38], dot immunoblot [39], PCR [40, 41], stool culture on selective media [42, 43], and enzyme immunoassay (EIA) [44, 45]. Because of differences between the hamster model and humans, it was originally believed that toxin A was important in human disease and many early EIA tests only detected toxin A, leading to false negative tests [46, 47].

3. Clinical signs and symptoms

Watery diarrhea with a distinct odor is usually the hallmark of *C. difficile* infection. Mild disease consists of crampy, watery diarrhea without systemic symptoms. This cohort constitutes 70% of patients with *C. difficile* infection as only about 30% of patients with *C. difficile* infection are febrile and 50% have a leukocytosis [48]. In severe disease, fecal leukocytes are generally high and diagnosis can be confirmed with endoscopy demonstrating pseudomembranous colitis. Other signs and symptoms of severe disease include abdominal pain, leukocytosis, and fever or other systemic symptoms. Leukocytosis is directly correlated with the severity of the disease. The elevation in white blood cell count can be as marginal as 15,000 cells/mL or as high as 50,000 cells/mL. Complications may include paralytic ileus, toxic megacolon, or other life threatening conditions. Postoperative patients and other patients with altered gastrointestinal motility may have pseudomembranous colitis without diarrhea secondary to ileus. Computed tomography is useful with characteristics of colitis readily seen on imaging that may include colonic wall thickening and associated ascites or toxic megacolon [21, 49].

Patients in the ICU tend to demonstrate the same spectrum of disease signs and symptoms as other infected persons. However, due to their illnesses, comorbidities weakened immune system and reduced ability to heal; the progression of the disease may advance more rapidly. Therefore, continual assessment of diarrhea and other symptoms of *C. difficile* infection is necessary as the severity may progress and further impact the already impaired and critical status of the patient in the ICU.

4. Risk factors for *Clostridium difficile* infection

Risk factors for *C. difficile* infection fall under three categories. First category includes disruptions of the endogenous intestinal flora, perturbations of the mucosa, or immunomodulation by exogenous factors that can occur as a result of medications, procedures, or radiation therapy. Most hospitalized patients with *C. difficile* infection have been exposed to antibiotics within the past 30 days. More recently, it has been noted that medications that suppress gastric acid, including proton-pump inhibitors and H_2-receptor blockers, increase risk of *C. difficile* infection, though study results are not uniform and the mechanism is not known [50–54]. For patients with primary or recurrent *C. difficile* infection, consideration should be given to discontinuation of gastric acid suppressants unless the patient's risk for GI bleeding outweighs the risk of *C. difficile* infection treatment failure. Chemotherapy, medications for autoimmune conditions, transplant medications, and radiation of the bowel increase the risk of *C. difficile* infection by disrupting the normal intestinal mucosal barrier and inhibiting the body's immunodefenses. Nasogastric tubes and enemas, presumably because of alteration of the normal flora and/or pH, increase patients' risk of *C. difficile* infection [55].

The second category of risk factors relates to how patients contract *C. difficile* infection. The most common method is by coming in contact with *C. difficile* spores from the hands of health care workers. Risk of contracting *C. difficile* infection is directly related to length of stay (LOS). Patients with longer LOS have multifactorial risk factors that include more severe illnesses that have a higher likelihood that they will require antibiotics and more prolonged exposure and interactions with health care workers [56, 57]. A patient's risk of contracting *C. difficile* infection is also related to *C. difficile* infection pressure that relates to the number of patients with *C. difficile* infection in a given care area [58]. Certain *C. difficile* strains, including the epidemic BI/NAP1/027 strain, have been isolated from prepared foods, pets, and from livestock [59–61].

The third category of risk factors relates to innate host susceptibility. Age >65 years is related to both an increased risk of primary *C. difficile* infection as well as an increased risk of more severe *C. difficile* infection. It is not known whether this is related to immune senescence, more frequent antibiotic usage, or increased comorbidities. The four comorbidities that place patients at greatest risk are sepsis, pneumonia, urinary tract infections, and skin infections—all of which generally require antibiotics for treatment. Patients hospitalized with higher numbers of conditions are more likely to contract *C. difficile* infection than patients with fewer conditions [48]. More recently, it has been noted that peripartum women and infants also appear to be at increased risk for *C. difficile* infection, including severe *C. difficile* infection related to the epidemic BI/NAP1/027 strain [62, 63]. Patients with inflammatory bowel disease (IBD) are more susceptible to *C. difficile* infection for reasons that are likely multifactorial, including antibiotic exposure, altered gut mucosal integrity, and immunosuppressive therapy. Patients with *C. difficile* infection superimposed on a flare of IBD are at risk for a particularly fulminant course. Because of altered gut physiology, patients with IBD may not develop pseudomembranes and may have a complicated diagnosis. Additionally, administration of glucocorticoids to treat the IBD exacerbation may predispose to *C. difficile* infection progression [64, 65]. Studies have shown that patients with HIV/AIDS or chronic kidney disease requiring hemodialysis are also at increased risk of *C. difficile* infection, possibly due to increased health care worker exposure or less robust immune response [66, 67].

5. Diagnosis

In the modern era, multiple tools have been developed to identify and detect *C. difficile* to include cultures, polymerase chain reaction (PCR), and enzyme immunoassays (EIA). Culturing *C. difficile* is difficult due to the strict anaerobic nature of the organism and the oxygen sensitivity that can kill the living organism. Utilizing an anaerobic chamber with a composition of 5% CO_2, 10% H_2, and 85% N_2 along with an air lock, has allowed the culturing, preservation, and storage of the living organism and spores [35, 43]. Once the organism has been cultured, PCR or EIA techniques can be utilized to detect toxin within the culture. These same techniques can be used independently of a culture to detect toxin within the stool sample. PCR has been successfully used since 1985 to amplify the 8.1 kilo base-pairs of the toxin A gene. Using 35 cycles of alternating 95–55°C temperatures and a Southern blot to isolate the 252 base-pair DNA fragment, PCR has become easy and commonplace for identification of the toxins [40, 41]. EIA has similarly been used since the early 80s for detection of both toxin A and B. The early tests were able to detect levels of toxin to 0.1 ng using a double sandwich microtiter plate with specificities of 98.6% and 100% for toxin A and toxin B, respectively [42–45]. More recently, glutamate dehydrogenase-immunoassay has been used as an initial screening tool with a chemiluminescent toxin-immunoassay for confirmation of both toxins A and B. The combined two-step process has a sensitivity and specificity of 100% [68]. The premise of the EIA tests is that antibodies to the toxins are attached to a plate. When the toxins pass over the antibodies, they become bound. A second preparation of antibodies with a marker attached to them is then added and a device to detect the markers allows for quantitative evaluation of the toxins present.

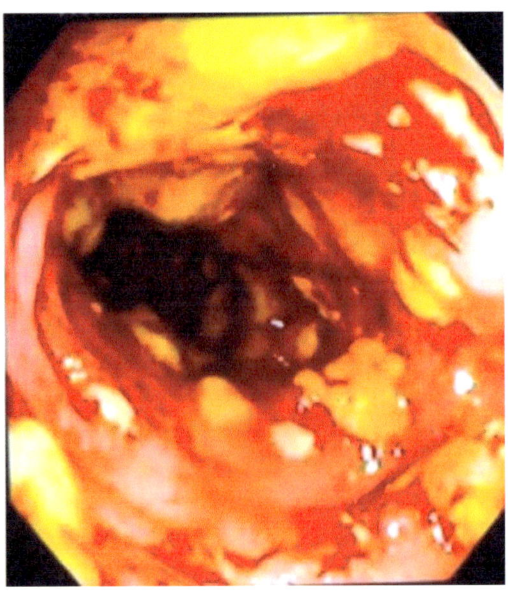

Figure 2. *Clostridium difficile* associated pseudomembranous colitis. Credit: North American Society for Pediatric Gastroenterology, Hepatology and Nutrition.

In addition to laboratory tests, computed tomography is useful to evaluate for toxic megacolon and colitis. When there is high clinical suspicion yet laboratory diagnostic tests have yielded negative results, the definitive test is colonoscopy. The appearance of pseudomembranes in the clinical setting of *C. difficile* infection is confirmatory for the diagnosis (**Figure 2**). **Table 1** displays the various current diagnostic modalities.

Intensivists should be familiar with the tests offered in their institution and be able to interpret the laboratory results in the context of clinical presentation. When clinical suspicion for *C. difficile* infection is high, the intensivist should initiate empiric therapy for *C. difficile* infection regardless of the diagnostic test results [48].

Test	Detection	Time	Usefulness
Culture	*Clostridium difficile*	34 days	Nonspecific and not useful for detection of toxins
Culture-toxins	Toxigenic *Clostridium difficile*	3–4 days	Must have initial growth from culture prior to testing for toxins
Cytotoxin	Toxin B	2–3 days	Costly and time-consuming. Results not immediately available
EIA toxin A & B	Toxin A & B	2–3 hours	Very quick but not sensitive. Need 3 specimens for increased sensitivity
EIA GDH	*Clostridium difficile*	2–3 hours	Screening test. Detects presence of bacteria but not specific
Toxin B gene	Toxigenic *Clostridium difficile*	2–3 hours	Very sensitive for detection of toxigenic strains of *Clostridium difficile* using PCR
Colonoscopy	Pseudomembranes	<1 hour	Very specific and sensitive for the detection of pseudomembranes
CT scan	Colitis	<1 hour	Very sensitive for colitis but not specific for *Clostridium difficile* infections

EIA, Enzyme Immunoassay; GDH, glutamine dehydrogenase; PCR, polymerase chain reaction; CT, computed tomography

Table 1. Diagnostic modalities for the identification of *Clostridium difficile* in the ICU.

6. Treatment

Once diagnosed, the first line of treatment is to discontinue implicated antibiotics, gastric acid suppression medications, and antiperistaltic medications, including narcotics and antimotility agents. Reduced peristalsis may prolong toxin exposure to the colonic mucosa [7]. Unfortunately, a large proportion of patients who develop *C. difficile* infection have documented infections that require treatment with antibiotics, and in the ICU setting, this proportion may reach 60% [69]. When it is not possible to stop antibiotic therapy, it is best to tailor coverage to more narrow spectrum agents once cultures and sensitivities are available. It is recommended to transition as soon as possible to β-lactams, macrolides, aminoglycosides, anti-staphylococcal drugs, tetracyclines, and other agents that have a lower likelihood of causing *C. difficile* infection [70].

Oral vancomycin is the only agent currently approved for treatment of *C. difficile* infection, although metronidazole in both oral and intravenous forms has been shown to be effective in treating *C. difficile* infection. Intravenous vancomycin has not been shown to be effective. Metronidazole has become the preferred agent for initial treatment of *C. difficile* infection because of lower cost [71, 72] and because of concerns over the possibility of increased development of vancomycin-resistant enterococcus [73, 74]. Metronidazole should be considered first-line therapy for mild to moderate *C. difficile* infection; however, it does have disadvantages compared to oral vancomycin. In a study involving 207 patients with *C. difficile* infection, 22% of patients remained symptomatic after 10 day therapy with metronidazole and 27% developed a relapse [75]. In a separate randomized trial involving 150 patients, the cure rate for metronidazole was only 76% compared with a 97% cure rate after treatment with vancomycin for the treatment of severe *C. difficile* infection [49]. Based on these studies and other data, oral vancomycin should be considered superior in the treatment of severe infections when GI motility is intact (**Table 2**) [49]. The pharmacology of oral vancomycin lends itself to being more effective as it is not absorbed by the GI tract and reaches the colon in high concentrations. The usual dosing regimen of 125 mg achieves levels of vancomycin 500–1000 times the minimal inhibitory concentration (MIC) of 90% of *C. difficile* in stool [48].

If the patient has ileus or severe pseudomembranous colitis and medication cannot be given orally, the use of rectal instillation of vancomycin solutions is supported by case reports [70, 76, 77]. The addition of intravenous metronidazole to either oral or intracolonic vancomycin in severely ill patients with ileus has been described, although this approach has not been adequately studied [78, 79].

Fidaxomicin is the first member in a new class of narrow spectrum macrocyclic antibiotics that are enterally administered and minimally absorbed in the GI tract. Having excellent *in vitro* and *in vivo* activity against *C. difficile*, including NAP1/BI/027 strains, and, while exhibiting limited activity *in vitro* and *in vivo* against components of the normal gut flora, fidaxomicin is an excellent candidate for replacing other agents in the treatment of *C. difficile* infections [80]. In a prospective, multicenter, double-blind, randomized, parallel-group trial involving 596

Severity	Preference	Medications
Mild CDI	1st line Alternate (PO) Alternate (IV)	Metronidazole: 500 mg PO every 8 hours Vancomycin: 125 mg PO every 6 hours or Fidaxomicin: 200 mg PO every 12 hours Metronidazole: 500 mg IV every 8 hours
Severe CDI	1st line Alternate (IV)	Vancomycin: 125 mg PO every 6 hours Metronidazole: 500 mg IV every 8 hours
Life-threatening CDI	1st line	Vancomycin: 500 mg every 6 hours via NGT or by enema plus Metronidazole: 500 mg IV every 8 hours
Relapsed CDI	1st line	Treatment based on severity as above

Table 2. Treatment modalities for *Clostridium difficile* infections.

patients, of which 287 received fidaxomicin and 309 received vancomycin, 88.2% of patients in the fidaxomicin group and 85.8% of those in the vancomycin group met the criteria for clinical cure. In addition, treatment with fidaxomicin was associated with a significantly lower rate of recurrence than was treatment with vancomycin (15.4 vs. 25.3%). More studies are warranted but results are promising [49, 81, 82].

Regardless of the type of medication, early treatment has been supported as the most effective pharmacologic treatment. A study by Zahar et al. conducted in three French ICUs has demonstrated that early treatment of ICU-acquired *C. difficile* infection results in mortality rates consistent with a control population of other ICU patients that have developed diarrhea that is not *C. difficile* infection associated. Treatment was initiated within 24 hours of onset and consisted of either metronidazole or oral vancomycin. The study involved 5,260 patients with an incidence of ICU-acquired diarrhea of 9.7%. All those with diarrhea were tested for *C. difficile* infection and 13.5% of those tested had confirmed toxin A or B by EIA and further confirmation by culture. None of the positive cultures produced any of the hypervirulent NAP1/027 strains seen in North American outbreaks. Overall mortality of ICU-acquired *C. difficile* infection was not independently associated with higher mortality rates compared to other patients with diarrhea in the ICU when matched for severity of illness, comorbidities, or complications occurring in the ICU. However, both the overall hospital stay and ICU stay was prolonged in the ICU-acquired *C. difficile* infection patients when compared to ICU patients as a whole (median 4 vs. 20 days) and ICU patients with diarrhea not associated with *C. difficile* infection (median 17 vs. 20 days). Despite these prolonged median stays, analysis did not demonstrate a statistically significant difference in length of stay with an estimated increase in overall ICU stay of 6.3 days ± 4.3, $p = 0.14$ compared to other ICU patients with diarrhea [83].

Microbial therapy with fecal transplantation can be accomplished with instillation of liquid preparations of stool from healthy donors. This method has proven successful for treating recurrent *C. difficile* infection in 70–100% of cases [84]. Probiotics may prevent attachment of *C. difficile* to epithelial cells and can reduce the incidence of *C. difficile* infection. *Saccharomyces boulardii* in particular has proven to be effective [49] whereas the use of *Lactobacillus* with conventional antibiotic therapy has shown mixed results including some studies showing no benefit in the treatment of *C. difficile* infection in several randomized controlled trials [85–88].

Use of anion exchange resins, such as cholestyramine and colestipol, with the hope of binding *C. difficile* cytotoxins in the treatment of *C. difficile* infection, has not only been shown to be effective [89, 90], but also carries the theoretical risk of binding intraluminal vancomycin, thus resulting in subtherapeutic vancomycin levels [91]. Intravenous immunoglobulins have been suggested for treatment of *C. difficile* infection but due to an insufficient evidence base and conflicting data, its use cannot be generally recommended until further studies have been conducted [92, 93]. Subtotal colectomy should be considered if there is no response to medical therapy within 3–4 days or if the patient remains seriously ill to avoid complications such as colonic perforation and sepsis [7].

7. Treatment failure and relapse

Patient characteristics that predispose to metronidazole failure include low serum albumin, continued exposure to the inciting antibiotic, and residence in the ICU [94, 95]. Particularly worrisome and concerning is the finding that relapsing or recurrent infections occur in up to 30% of patients treated for *C. difficile* infection whether the initial treatment was metronidazole or vancomycin [96]. This could be due to reinfection with the same endogenous strain or from a different strain acquired exogenously. Patients that had an initial infection followed by reinfection have a 50–65% chance of further repeated episodes. A metaanalysis by Garey et al. found that reexposure to antimicrobials, gastric acid suppression, and older age are all associated with an increased risk of recurrent *C. difficile* infection [97]. Patients that have three or more episodes of *C. difficile* infection, considered to be multiple *C. difficile* infection recurrence, are best treated with a tapered regimen of oral vancomycin. The initial dose of vancomycin administered is at the usual 125 mg by mouth four times a day for 10–14 days but then one dose per day is removed one week at a time until the patient is taking one dose every 2–3 days. The rationale for this regimen is that as the doses are spaced out, the colonic flora has time to regenerate [48].

8. Generating optimal colonic flora for risk reduction

There is an urgent need for alternative means of preventing and treating *C. difficile* infection in high-risk individuals. Metagenomics have improved our understanding of the "colonization resistance barrier" and how this could be optimized. The "colonization resistance barrier" in the normal healthy colon consists of high microbial diversity, substrate/area competition, immune response modulation and short-chain fatty acid (SCFA) production [16, 98]. These factors are often missing in the elderly. Decreased pH, oxidation-reduction potentials, and higher concentrations of short-chain fatty acids have been suggested to inhibit *C. difficile* growth and toxin production throughout *in vitro* and *in vivo* studies. There is, therefore, evidence in support of a colonization resistance barrier against *C. difficile* infection [16, 98].

For instance, *in vitro, Bifidobacterium longum* and *Bifidobacterium breve* have been show to significantly reduce the growth of the toxigenic strain *C. difficile* LMG21717 [99]. In a randomized, placebo-controlled, double-blind trial at a long-term elderly care facility, the effectiveness of a *Lactobacillus casei* strain Shirota (LcS) infused beverage was demonstrated by altering *Clostridium* infection rates among the residents. Daily consumption of the beverage resulted in a significantly lower incidence of fever and improved bowel movements. When compared to a resident control group drinking a placebo beverage, stool studies from the experimental LcS group showed significantly higher number of both *Bifidobacterium* and *Lactobacillus* ($p < 0.01$), significantly lower number of destructive bacteria such as *C. difficile* ($p < 0.05$), and a higher fecal acetic acid concentration. This study was also conducted among the facility's staff and a significant difference in the intestinal microbiota, fecal acetic acid, and pH was also observed between the LcS and placebo groups [100].

http://dx.doi.org/10.5772/intechopen.69212

There is some evidence to support that plant based diets may reduce the number of pathobionts such as *C. difficile* and increase the number of protective species such as Lactobacillus [100–103]. Altered flora with resulting altered bile metabolism within the gut by flora favored by plant-based diets have implications in colonocyte protection [102]. Intestinal microbiota are able to produce short chain fatty acids (SCFA), such as acetate, propionate, and butyrate, through metabolism of dietary fiber. These SCFA have been shown to be colonocyte protective. A strong positive correlation has been found between *Faecalibacterium prausnitzii* and butyrate production in the gastrointestinal tract, suggesting that this species may be associated with higher fiber intake and reduced risk, not only for *C. difficile* infection, but also for other common comorbidities in the elderly including cardiovascular disease, colon cancer, diabetes, and obesity [104]. A move toward a diet that decreases risk for contracting *C. difficile* infection should be encouraged, not only in the elderly, but also generally, because of the broad implications.

9. Prevention of hospital spread

Disinfectant products based on quaternary ammonium compounds, commonly used to clean patient rooms, are not sporicidal. Therefore, using sporicidal hypochlorite-based disinfectants on surfaces is recommended. However, use of antisporicidal agents outside an outbreak is not associated with lower rates of *C. difficile* infection [105, 106].

Hand hygiene is the most important preventive measure to reduce transmission of *C. difficile* spores. Soap and water has been demonstrated to be superior to alcohol based hand rubs and other forms of hand sanitation with regard to transmission by healthcare workers [21, 107]. Hospital hygiene hand protocols should be followed assiduously at all times. Other precautions that should be utilized include isolation of the patient, barrier precautions, and use of chlorine based chemical wipes [107]. These precautions should not be lifted based on stool studies as there are no diagnostic methods to determine response to treatment. Rather, the decision should be made on clinical signs and symptoms with resolution of diarrhea, fevers, and leukocytosis. A strong antibiotic stewardship program is essential to limit the use of antibiotics that may cause *C. difficile* infection and is generally a good principle to follow. It has been demonstrated that up to 25% of antibiotic administration is not indicated, even in the ICU [108].

10. *Clostridium difficile* infection in the intensive care unit

Diarrhea is a common problem in the ICU affecting up to 40% of patients admitted. Severely burned patients may have an incidence of greater than 90% [109, 110]. Enteral tube feeding is the most common cause of diarrhea in the ICU; other causes include hypoalbuminemia, intestinal ischemia, and medications. *C. difficile* infection is the most common infectious cause of diarrhea in the ICU [111, 112]. The severity of *C. difficile* infection is increasing which is possibly related to the emergence of more virulent strains such as the BI/NAP1/027 strain, prompting more admissions to the ICU for management of *C. difficile* infection related complications [113].

In a systematic review and metaanalysis of 22 published studies from 1983 to 2015 that included 80,835 ICU patients, the effects of *C. difficile* infection on morbidity and mortality were investigated. Karanika et al. found that prevalence of *C. difficile* infection among ICU patients was 2% but 5-fold greater in those patients with diarrhea (11%). Those patients that were diagnosed with *C. difficile* infection had a 25% incidence of the severe form of the disease and diagnosed with pseudomembranous colitis. ICU mortality was not significantly different between the group with *C. difficile* infection and the nonC. *difficile* infection group based on seven studies that enrolled a combined 12,165 patients. However, the overall hospital mortality between those same groups was significantly increased in the *C. difficile* infection group with 32% mortality compared to 24% (p = 0.03). Similarly, length of ICU and hospital stay among *C. difficile* infection patients was longer when compared to nonC. *difficile* infection patients. Based on five studies with over 10,000 patients, *C. difficile* infection patients had an average ICU stay of 24 days and overall hospital stay of 50 days compared to 19 days and 30 days, respectively, for the nonC. *difficile* infection group (p = 0.001) [114].

Even though only 3% of patients with *C. difficile* infection require subtotal colectomy for fulminant *C. difficile* colitis, 20% of ICU patients with severe *C. difficile* infection will still require partial colectomy or diversion [115, 116]. Colectomy in this setting is associated with a 50% mortality [90]. Mortality rates are lower when surgical intervention is undertaken within 48 hours of lack of response to medical therapy [117]. During NAP1/027 outbreaks, patients with age >65 years, leukocytosis and elevated lactate appear to benefit the most from early colectomy [118].

In a series of 29 patients with severe or severe/complicated *C. difficile* infection refractory to oral vancomycin ± rectal vancomycin and intravenous metronidazole therapy who underwent fecal microbiota transplantation (FMT) plus continued vancomycin, overall treatment response was 93% (27/29), including 100% (10/10) for severe *C. difficile* infection and 89% (17/19) for severe/complicated *C. difficile* infection. A single FMT was performed in 62%, two FMTs were performed in 31%, and three FMTs in 7% of patients. Continued use of non *C. difficile* infection antibiotics predicted repeat FMT. Thirty-day all-cause mortality after FMT was 7%. Of the two patients who died within 30 days, one underwent colectomy and succumbed to sepsis; the other died from septic shock related to *C. difficile* infection [84]. Further research into the use of FMT combined with continued vancomycin is needed.

11. Modern outbreaks

In 2003, a major outbreak of *C. difficile* occurred in Quebec, Canada and was identified as ribotype 027, strain BI/NAP1. This strain has been identified in >50% of all isolates from hospitals in Europe and North America [4, 10, 20]. Prior to the 2003 outbreak, this strain only accounted for 14 of over 6000 (<0.02%) typed strains collected from U.S. cases during the period of 1984 to 1993. Following the 2003 outbreak in Canada, 96 of 187 (51%) strains tested positive for 027 in eight U.S. outbreaks [119].

The BI/NAP1/027 strain belongs to a hypervirulent group of strains along with types 001, 017, and 078. In particular, the binary toxin produced by 027 was not seen previously. It is thought to be synergistic with the production of toxin A and B. Strain BI/NAP1/027 was found to be highly resistant to fluoroquinolone classes of antibiotics and was also found to produce 16-fold higher concentrations of toxin A and 23-fold higher concentrations of toxin B than less virulent toxinotype 0 strains. The binary toxin has been associated with more severe diarrhea when combined with toxin A and B. When produced alone, binary toxin does not appear to produce disease [3, 10] but does appear to be a marker of both *C. difficile* infection severity and recurrence [120]. The emergence is generally believed to be related to fluoroquinolone exposure though not to the particular type of fluoroquinolone [121, 122].

12. Conclusions

C. difficile is a very diverse group of toxin producing organisms. Newer technologies have allowed the identification of numerous toxinotypes and ribotypes with varying virulence factors and toxin production. Multiple lineages contain hypervirulent strains. The large degree of horizontal gene transfer through transposons, bacteriophages, and homologous recombination has dispersed genetic material and pathogenic properties among different strains.

The increased prevalence of ribotypes 027, 017, and 078 may be solely due to population expansion over the last decade or due to a nosocomial enrichment of the proper environment and conditions for the expansion and transference of these virulent strains. The sudden rise may also be related to the delay in purifying selection pressures seen in the more recently diverging lineages. However, a more likely explanation for increasing incidence is the right combination of elderly patients in a contaminated environment with antibiotic and acid suppression medications. Given the high incidence of colonized guts in the hospitalized pediatric population (70%), the hospitalized adult population (25%), the animal kingdom (40%), and the natural environment (50%), reducing exposure is near impossible [14].

The high virulence, along with a highly mobile genome capable of antibiotic resistance, has prompted further research in the development of vaccinations. Sanofi-Aventis is currently undergoing trials with a vaccine containing formalin-inactivated toxins A and B. To date, 100 healthy subjects have been exposed to the vaccine without any serious side effects [123].

The hardiness of *C. difficile* spores and the ease with which this bacterium alters its genome has allowed it to flourish and survive among a variety of hosts and reservoirs. More virulent strains are a real possibility given the mobility of code sequencing regions within the genome. As the population continues to age and makes an increasingly stronger presence throughout the healthcare system, especially in the ICU, *C. difficile* will continue to plague patients and healthcare providers until further measures are discovered to control transmission. The increased burden will stress the current resources and facilities financially, geographically, and the pool of available care takers. To date, the best treatment modalities include eliminating the implicated antibiotics, early initiation of oral vancomycin and metronidazole, and strict infection-control engineering to prevent the initial infection.

Author details

William C. Sherman, Chris Lewis, Jong O. Lee and David N. Herndon*

*Address all correspondence to: dherndon@utmb.edu

Department of Surgery, University of Texas Medical Branch – Galveston, Galveston, Texas, United States of America

References

[1] Hall IC, O'Toole E. Intestinal flora in newborn infants with a description of a new pathogenic anaerobe, *Bacillus difficilis*. American Journal of Diseases of Children. 1935;**49**:390-402

[2] Finney JMT. Gastroenterology for cicatrizing ulcer of the pylorus. Bulletin of the Johns Hopkins Hospital. 1893;**4**:53-55

[3] Cecil JA. *Clostridium difficile*: Changing epidemiology, treatment and infection prevention measures. Current Infectious Disease Reports. 2012;**14**(6):612-619

[4] He M, Sebaihia M, et al. Evolutionary dynamics of *Clostridium difficile* over short and long time scales. Proceedings of the National Academy of Sciences. 2010;**107**(16):7527-7532

[5] Carrico RM, Archibald LK, Bryant K, et al. Elimination Guide Series. Washington, DC: Association for Professionals in Infection Control and Epidemiology. APIC. Guide for Elimination of *Clostridium difficile* in Healthcare Settings; 2008. pp. 1-66

[6] Nearly Half a Million Americans Suffered from *Clostridium difficile* Infections in a Single Year [Internet]. 2015. Available from: https://www.cdc.gov/media/releases/2015/p0225-clostridium-difficile.html [Accessed: March 2, 2017]

[7] Bartlett JG. Clinical Practice. Antibiotic-associated diarrhea. The New England Journal of Medicine. 2002;**346**:334-339

[8] Chang D, Brouse S, Kelly K. Retrospective review of *Clostridium difficile* infections in the medical intensive care unit. Critical Care Medicine. 2011;**39**(12):163

[9] Weinstein RA. Intensive care unit environments and the fecal patina: a simple problem? Critical Care Medicine. 2012;**40**(4):1333-1334

[10] Association for Professionals in Infection Control and Epidemiology, Inc. APIC. Guide to the Elimination of *Clostridium difficile* in Healthcare Settings. 2008 November

[11] Voth DE, Ballard JD. *Clostridium difficile* Toxins: Mechanism of action and role in disease. Clinical Microbiology Reviews. 2005;**18**(2):247-263

[12] von Eichel-Streiber C, Laufenberg-Feldmann R, Sartingen S, Schulze J, Sauerborn M. Comparative sequence analysis of the *Clostridium difficile* toxins A and B. Molecular & General Genetics. 1992;**233**:260-268

[13] Riegler M, Sedivy R, Pothoulakis C, et al. *Clostridium difficile* toxin B is more potent than toxin A in damaging human colonic epithelium *in vitro*. The Journal of Clinical Investigation. 1995;**95**:2004-2011

[14] Carroll KC, Bartlett JG. Biology of *Clostridium difficile*: Implications for epidemiology and diagnosis. Annual Review Microbiology. 2011;**65**:501-521

[15] O'Brien JA, Lahue BJ, Caro JJ, Davidson DM. The emerging infectious challenge of *Clostridium difficile*-associated disease in Massachusetts hospitals: Clinical and economic consequences. Infection Control and Hospital Epidemiology. 2007;**28**(11):1219-1227

[16] Yuille S, Mackay WG, Morrison DJ, Tedford MC. Optimising gut colonisation resistance against *Clostridium difficile* infection. European Journal of Clinical Microbiology & Infectious Diseases. 2015;**34**(11):2161-2166

[17] Aseri M, Schroeder T, Kramer J, Kackula R. Gastric acid suppression by proton pump inhibitors as a risk factor for *Clostridium difficile*-associated diarrhea in hospitalized patients. The American Journal of Gastroenterology. 2008;**103**:2308-2313

[18] Ebright JR, Fekety R, Silva J, Wilson KH. Evaluation of eight cephalosporins in hamster colitis model. Antimicrob Agents Chemother. 1981;**19**:980-986

[19] Fekety R, Silva J, Toshniwal R, et al. Antibiotic-associated colitis: Effects of antibiotics on *Clostridium difficile* and the disease in hamsters. Reviews of Infectious Diseases. 1979;**1**(2):386-397

[20] Davies KA, Ashwin H, Longshaw CM, Burns DA, Davis GL, Wilcox MH; EUCLID study group. Diversity of *Clostridium difficile* PCR ribotypes in Europe: results from the European multicentre, prospective, biannual, point-prevalence study of *Clostridium difficile* infection in hospitalised patients with diarrhoea (EUCLID), 2012 and 2013. Euro Surveillance. 2016;**21**(29):1-11

[21] Cohen SH, Gerding DN, Johnson S, Kelly CP, Loo VG, et al. Clinical practice guidelines for *Clostridium difficile* infection in adults: 2010 update by the Society for Healthcare Epidemiology of America (SHEA) and the Infectious Diseases Society of America (IDSA). Infection Control and Hospital Epidemiology. 2010;**31**:431-455

[22] Altemeier WA, Hummel RP, Hill EO. Staphylococcal enterocolitis following antibiotic therapy. Annals of Surgery. 1963;**157**:847-858

[23] Hummel RP, Altemeier WA, Hill EO. Iatrogenic staphylococcal enterocolitis. Annals of Surgery. 1964;**160**:551-560

[24] Khan MY, Hall WH. Staphylococcal enterocolitis—treatment with oral vancomycin. Annals of Internal Medicine. 1966;**65**:1-8

[25] Tedesco FJ, Corless JK, Brownstein RE. Rectal sparing in antibiotic-associated pseudomembranous colitis: A prospective study. Gastroenterology. 1982;**83**:1259-1260

[26] Green RH. The association of viral activation with penicillin toxicity in guinea pigs and hamsters. The Yale Journal of Biology and Medicine. 1974;**47**:166-181

[27] Keighley MR, Burdon DW, Arabi Y, et al. Randomised controlled trial of vancomycin for pseudomembranous colitis and postoperative diarrhoea. British Medical Journal. 1978;**2**:1667-1669

[28] Lusk RH, Fekety Jr FR, Silva Jr J, et al. Gastrointestinal side effects of clindamycin and ampicillin therapy. The Journal of Infectious Diseases. 1977;**135**(Suppl):111-119

[29] Bartlett JG, Onderdonk AB, Cisneros RL, Kasper DL. Clindamycin-associated colitis due to a toxin-producing species of Clostridium in hamsters. The Journal of Infectious Diseases. 1977;**136**:701-705

[30] Taylor NS, Thorne GM, Bartlett JG. Comparison of two toxins produced by *Clostridium difficile*. Infection and Immunity. 1981;**34**:1036-1043

[31] Gilligan PH, McCarthy LR, Genta VM. Relative frequency of *Clostridium difficile* in patients with diarrheal disease. Journal of Clinical Microbiology. 1981;**14**:26-31

[32] Hirschhorn LR, Trnka Y, Onderdonk A, Lee ML, Platt R. Epidemiology of community-acquired *Clostridium difficile* -associated diarrhea. The Journal of Infectious Diseases. 1994;**169**:127-133

[33] Keighley MR. Antibiotic-associated pseudomembranous colitis: Pathogenesis andmanagement. Drugs. 1980;**20**:49-56

[34] Burbige EJ, Radigan JJ. Antibiotic-associated colitis with normal-appearing rectum. Diseases of the Colon and Rectum. 1981;**24**:198-200

[35] Bartlett J. Historical perspectives on studies of *Clostridium difficile* and *C. difficile* infection. Clinical Infectious Diseases. 2008;**46**(1 Suppl):S4-S11

[36] Shahrabadi MS, Bryan LE, Gaffney D, Coderre SE, Gordon R, Pai CH. Latex agglutination test for detection of *Clostridium difficile* toxin in stool samples. Journal of Clinical Microbiology. 1984;**20**:339-341

[37] Lyerly DM, Ball DW, Toth J, Wilkins TD. Characterization of cross-reactive proteinsdetected by Culturette Brand Rapid Latex Test for *Clostridium difficile.* Journal of Clinical Microbiology. 1988;**26**:397-400

[38] Lyerly DM, Barroso LA, Wilkins TD. Identification of the latex test-reactive protein of *Clostridium difficile* as glutamate dehydrogenase. Journal of Clinical Microbiology. 1991;**29**:2639-2642

[39] Kurzynski TA, Kimball JL, Schultz DA, Schell RF. Evaluation of *C. diff.*-CUBE testfor detection of *Clostridium difficile* -associated diarrhea. Diagnostic Microbiology and Infectious Disease. 1992;**15**:493-498

[40] Kato N, Ou CY, Kato H, et al. Detection of toxigenic *Clostridium difficile* in stoolspecimens by the polymerase chain reaction. The Journal of Infectious Diseases. 1993;**167**:455-458

[41] Kuhl SJ, Tang YJ, Navarro L, Gumerlock PH, Silva Jr J. Diagnosis and monitoringof *Clostridium difficile* infections with the polymerase chain reaction. Clinical Infectious Diseases. 1993;**16**(4 Suppl):234-238

[42] Laughon BE, Viscidi RP, Gdovin SL, Yolken RH, Bartlett JG. Enzyme immunoassaysfor detection of *Clostridium difficile* toxins A and B in fecal specimens. Clinical Infectious Diseases. 1984;**149**:781-788

[43] Walker RC, Ruane PJ, Rosenblatt JE, et al. Comparison of culture, cytotoxicity assays, and enzyme-linked immunosorbent assay for toxin A and toxin B in the diagnosis of *Clostridium difficile* -related enteric disease. Diagnostic Microbiology and Infectious Disease. 1986;**5**:61-69

[44] DiPersio JR, Varga FJ, Conwell DL, Kraft JA, Kozak KJ, Willis DH. Development of a rapid enzyme immunoassay for *Clostridium difficile* toxin A and its use in the diagnosis of *C. difficile*-associated disease. Journal of Clinical Microbiology. 1991;**29**:2724-2730

[45] Woods GL, Iwen PC. Comparison of a dot immunobinding assay, latex agglutination, and cytotoxin assay for laboratory diagnosis of *Clostridium difficile*-associated diarrhea. Journal of Clinical Microbiology. 1990;**28**:855-857

[46] Limaye AP, Turgeon DK, Cookson BT, Fritsche TR. Pseudomembranous colitis caused by a toxin A−B+ strain of *Clostridium difficile*. Journal of Clinical Microbiology. 2000;**38**: 1696-1697

[47] Kato H, Kato N, Watanabe K, et al. Identification of toxin A-negative, toxin B-positive *Clostridium difficile* by PCR. Journal of Clinical Microbiology. 1998;**36**:2178-2182

[48] Bobo LD, Dubberke ER, Kollef M. *Clostridium difficile* in the ICU: The struggle continues. Chest. 2011;**140**(6):1643-1653

[49] Zar FA, Bakkanagari SR, Moorthi KM, Davis MB. A comparison of vancomycin and metronidazole for the treatment of *Clostridium difficile*-associated diarrhea, stratified by disease severity. Clinical Infectious Diseases. 2007;**45**:302-307

[50] Yearsley KA, Gilby LJ, Ramadas AV, Kubiak EM, Fone DL, Allison MC. Proton pump inhibitor therapy is a risk factor for *Clostridium difficile* associated diarrhoea. Alimentary Pharmacology & Therapeutics. 2006;**24**(4):613-619

[51] Dial S, Delaney JA, Barkun AN, Suissa S. Use of gastric acid suppressive agents and the risk of community-acquired *Clostridium difficile*-associated disease. The Journal of the American Medical Association. 2005;**294**(23):2989-2995

[52] Cadle RM, Mansouri MD, Logan N, Kudva DR, Musher DM. Association of proton pumpinhibitors with outcomes in *Clostridium difficile* colitis. American Journal of Health System Pharmacy. 2007;**64**(22):2359-2363

[53] Linsky A, Gupta K, Lawler EV, Fonda JR, Hermos JA. Proton pump inhibitors and risk for recurrent *Clostridium difficile* infection. Archives of Internal Medicine. 2010;**170**(9):772-778

[54] Howell MD, Novack V, Grgurich P, et al. Iatrogenic gastric acid suppression and the risk of nosocomial *Clostridium difficile* infection. Archives of Internal Medicine. 2010;**170**(9):784-790

[55] Kurd MF, Pulido L, Joshi A, Purtill JJ, Parvizi J. *Clostridium difficile* infection after total joint arthroplasty: Who is at risk? The Journal of Arthroplasty. 2008;**23**(6):839-842

[56] Sunenshine RH, McDonald LC. *Clostridium difficile*-associated disease: New challenges from an established pathogen. Cleveland Clinic Journal of Medicine. 2006;**73**(2):187-197

[57] Modena S, Bearelly D, Swartz K, Friedenberg FK. *Clostridium difficile* among hospitalized patients receiving antibiotics: a case control study. Infection Control and Hospital Epidemiology. 2005;**26**(8):685-690

[58] Dubberke ER, Yan Y, Reske KA, et al. Development and validation of a *Clostridium difficile* infection risk prediction model. Infection Control and Hospital Epidemiology. 2011;**32**(4):360-366

[59] Rupnik M, Songer JG. *Clostridium difficile*: Its potential as a source of foodborne disease. Advances in Food and Nutrition Research. 2010;**60C**:53-66

[60] Jhung MA, Thompson AD, Killgore GE, et al. Toxinotype V *Clostridium difficile* in humans and food animals. Emerging Infectious Diseases. 2008;**14**(7):1039-1045

[61] Yaeger MJ, Kinyon JM, Glenn Songer J. A prospective, case control study evaluating the association between *Clostridium difficile* toxins in the colon of neonatal swine and gross and microscopic lesions. Journal of Veterinary Diagnostic Investigation. 2007;**19**(1):52-59

[62] Rouphael NG, O'Donnell JA, Bhatnagar J, et al. *Clostridium difficile*-associated diarrhea: an emerging threat to pregnant women. American Journal of Obstetrics and Gynecology. 2008;**198**(6):635

[63] Pépin J, Saheb N, Coulombe MA, et al. Emergence of fluoroquinolones as the predominant risk factor for *Clostridium difficile*-associated diarrhea: a cohort study during an epidemic in Quebec. Clinical Infectious Diseases. 2005;**41**(9):1254-1260

[64] Issa M, Vijayapal A, Graham MB, et al. Impact of *Clostridium difficile* on inflammatory bowel disease. Clinical Gastroenterology and Hepatology. 2007;**5**(3):345-351

[65] Adams SD, Mercer DW. Fulminant *Clostridium difficile* colitis. Current Opinion in Critical Care. 2007;**13**(4):450-455

[66] Raines DL, Lopez FA. *Clostridium difficile* in non-HIV-immunocompromised patients and HIV-infected patients. Current Gastroenterology Reports. 2011;**13**(4):344-350

[67] Sanchez TH, Brooks JT, Sullivan PS, et al. Adult/Adolescent Spectrum of HIV Disease Study Group Bacterial diarrhea in persons with HIV infection, United States, 1992-2002. Clinical Infectious Diseases. 2005;**41**(11):1621-1627

[68] Makristathis A, Zeller I, Mitteregger D, Kundi M, Hirschl AM. Comprehensive evaluation of chemiluminescent immunoassays for the laboratory diagnosis of *Clostridium difficile* infection. European Journal of Clinical Microbiology & Infectious Diseases. 2017:1-7. [epub ahead of print]

[69] Marra AR, Edmond MB, Wenzel RP, et al. Hospital-acquired *Clostridium difficile*-associated disease in the intensive care unit setting: epidemiology, clinical course and outcome. BMC Infectious Diseases. 2007;**7**:42

[70] Malnick SD, Zimhony O. Treatment of *Clostridium difficile* associated diarrhea. The Annals of Pharmacotherapy. 2002;**36**:1767-1775

[71] Teasley DG, Gerding DN, Olson MM, et al. Prospective randomised trial of metronidazole versus vancomycin for *Clostridium difficile*-associated diarrhoea and colitis. Lancet. 1983;**2**:1043-1046

[72] Wenisch C, Parschalk B, Hasenhundl M, et al. Comparison of vancomycin, teicoplanin, metronidazole, and fusidic acid for the treatment of *Clostridium difficile*-associated diarrhea. Clinical Infectious Diseases. 1996;**22**:813-818

[73] AlNassir WN, Sethi AK, Li Y, et al. Both oral metronidazole and oral vancomycin promote persistent overgrowth of vancomycin-resistant enterococci during treatment of *Clostridium difficile* associated disease. Antimicrobial Agents and Chemotherapy. 2008;**52**:2403-2406

[74] Salgado CD, Giannetta ET, Farr BM. Failure to develop vancomycin-resistant Enterococcus with oral vancomycin treatment of *Clostridium difficile*. Infection Control and Hospital Epidemiology. 2004;**25**:413-417

[75] Musher DM, Aslam S, Logan N, et al. Relatively poor outcome after treatment of *Clostridium difficile* colitis with metronidazole. Clinical Infectious Diseases. 2005;**40**:1586-1590

[76] Apisarnthanarak A, Razavi B, Mundy LM. Adjunctive intracolonic vancomycin for severe *Clostridium difficile* colitis: Case series and review of the literature. Clinical Infectious Diseases. 2002;**35**:690-696

[77] Nathanson DR, Sheahan M, Chao L, et al. Intracolonic use of vancomycin for treatment of *Clostridium difficile* colitis in a patient with a diverted colon: report of a case. Diseases of the Colon and Rectum. 2001;**44**:1871-1872

[78] Friedenberg F, Fernandez A, Kaul V, et al. Intravenous metronidazole for the treatment of *Clostridium difficile* colitis. Diseases of the Colon and Rectum. 2001;**44**:1176-1180

[79] Johnson S, Peterson LR, Gerding DN. Intravenous metronidazole and *Clostridium difficile* associated diarrhea or colitis. The Journal of Infectious Diseases. 1989;**160**:1087-1088

[80] van Nispen tot Pannerden CMF, Verbon A, Kuipers E. Recurrent *Clostridium difficile* infection. What are the treatment options? Drugs. 2011;**71**:853-868

[81] Louie TJ, Miller MA, Mullane KM, Weiss K, Lentnek A, Golan Y, Gorbach S, Sears P, Shue YKFidaxomicin versus Vancomycin for *Clostridium difficile* Infection. The New England Journal of Medicine. 2011;**364**:422-431

[82] Spiceland CM, Khanna S, Pardi DS. Outcomes with fidaxomicin therapy in *Clostridium difficile* infection. Journal of Clinical Gastroenterology. 2016. Forthcoming 2017.

[83] Zahar et al. Outcome of ICU patients with *Clostridium difficile* infection. Critical Care. 2012;**16**(R215):1-10

[84] Fischer M, Sipe BW, Rogers NA, Cook GK, Robb BW, Vuppalanchi R, Rex DK. Faecal microbiota transplantation plus selected use of vancomycin for severe-complicated

Clostridium difficile infection: Description of a protocol with high success rate. Alimentary Pharmacology & Therapeutics. 2015;**42**(4):470-476

[85] Lawrence SJ, Korzenik JR, Mundy LM. Probiotics for recurrent *Clostridium difficile* disease. Journal of Medical Microbiology. 2005;**54**:905-906

[86] McFarland LV, Surawicz CM, Greenberg RN, et al. A randomized placebo-controlled trial of *Saccharomyces boulardii* in combination with standard antibiotics for *Clostridium difficile* disease. The Journal of the American Medical Association. 1994;**271**:1913-1918

[87] Surawicz CM, McFarland LV, Greenberg RN, et al. The search for a better treatment for recurrent *Clostridium difficile* disease: Use of high-dose vancomycin combined with *Saccharomyces boulardii*. Clinical Infectious Diseases. 2000;**31**:1012-1017

[88] Wullt M, Hagslatt ML, Odenholt I. *Lactobacillus plantarum* 299v for the treatment of recurrent *Clostridium difficile*-associated diarrhoea: a doubleblind, placebo-controlled trial. Scandinavian The Journal of Infectious Diseases. 2003;**35**:365-367

[89] Ariano RE, Zhanel GG, Harding GK. The role of anion-exchange resins in the treatment of antibiotic-associated pseudomembranous colitis. Canadian Medical Association journal. 1990;**142**:1049-1051

[90] Mogg GA, Burdon DW, Keighley M. Oral metronidazole in *Clostridium difficile* colitis. British Medical Journal. 1979;**2**:335

[91] Taylor NS, Bartlett JG. Binding of *Clostridium difficile* cytotoxin and vancomycin by anion-exchange resins. The Journal of Infectious Diseases. 1980;**141**:92-97

[92] Salcedo J, Keates S, Pothoulakis C, et al. Intravenous immunoglobulin therapy for severe *Clostridium difficile* colitis. Gut. 1997;**41**:366-370

[93] Juang P, Skledar SJ, Zgheib NK, et al. Clinical outcomes of intravenous immune globulin in severe *Clostridium difficile* associated diarrhea. American Journal of Infection Control. 2007;**35**:131-137

[94] Fernandez A, Anand G, Friedenberg F. Factors associated with failure of metronidazole in *Clostridium Difficile*-associated disease. Journal of Clinical Gastroenterology. 2004;**38**:414-418

[95] Nair S, Yadav D, Corpuz M, et al. *Clostridium difficile* colitis: Factors influencing treatment failure and relapse: A prospective evaluation. The American Journal of Gastroenterology. 1998;**93**:1873-1876

[96] Bouza E, Muñoz P, Alonso R. Clinical manifestations, treatment and control of infections caused by *Clostridium difficile*. Clinical Microbiology and Infection. 2005;**11**(4 Suppl):57-64

[97] Garey KW, Sethi S, Yadav Y, DuPont HL. Meta-analysis to assess risk factors for recurrent *Clostridium difficile* infection. The Journal of Hospital Infection. 2008;**70**(4):298-304

[98] Latorre M, Krishnareddy S, Freedberg DE. Microbiome as mediator: Do systemic infections start in the gut? World Journal of Gastroenterology. 2015;**21**(37):10487-10492

[99] Valdés-Varela L, Hernández-Barranco AM, Ruas-Madiedo P, Gueimonde M. Effect of *Bifidobacterium* upon *Clostridium difficile* growth and Toxicity when Co-cultured in different prebiotic substrates. Frontiers in Microbiology. 2016;**7**:738

[100] Nagata S, Asahara T, Wang C, Suyama Y, Chonan O, Takano K, Daibou M, Takahashi T, Nomoto K, Yamashiro Y. The effectiveness of *Lactobacillus* beverages in controlling infections among the residents of an aged care facility: A randomized Placebo-Controlled Double-Blind Trial. Annals of Nutrition & Metabolism. 2016;**68**(1):51-59

[101] Chung KT, Kuo CT, Chang FJ. Detection of lactobacilli and their interaction with clostridia in human gastrointestinal tracts and *in vitro*. Zhonghua Minguo Wei Sheng Wu Ji Mian Yi Xue Za Zhi. 1989;**22**(3):163-172

[102] Matijašić BB, Obermajer T, Lipoglavšek L, Grabnar I, Avguštin G, Rogelj I. Association of dietary type with fecal microbiota in vegetarians and omnivores in Slovenia. European Journal of Nutrition. 2014;**53**(4):1051-1064

[103] Kabeerdoss J, Devi RS, Mary RR, Ramakrishna BS. Faecal microbiota composition in vegetarians: comparison with omnivores in a cohort of young women in southern India. The British Journal of Nutrition. 2012;**108**(6):953-957

[104] Benus R, van der Werf TS, Welling GW, Judd PA, Taylor MA, Harmsen HJM, Whelan K. Association between *Faecalibacterium prausnitzii* and dietary fibre in colonic fermentation in healthy human subjects. The British Journal of Nutrition. 2010;**104**:693-700

[105] Shim JK, Johnson S, Samore MH, Bliss DZ, Gerding DN. Primary symptomless colonisation by *Clostridium difficile* and decreased risk of subsequent diarrhoea. Lancet. 1998;**351**(9103):633-636

[106] Mayfield JL, Leet T, Miller J, Mundy LM. Environmental control to reduce transmission of *Clostridium difficile*. Clinical Infectious Diseases. 2000;**31**(4):995-1000

[107] Cooper CC, Jump RL, Chopra T. Prevention of infection due to *Clostridium difficile*. Infectious Disease Clinics of North America. 2016;**30**(4):999-1012

[108] Lawrence KL, Kollef MH. Antimicrobial stewardship in the intensive care unit: Advances and obstacles. American Journal of Respiratory and Critical Care Medicine. 2009;**179**(6):434-438

[109] Kelly TW, Patrick MR, Hillman KM. Study of diarrhea in critically ill patients. Critical Care Medicine. 1983;**11**:7-9

[110] Thakkar K, Kien CL, Rosenblatt JI, Herndon D. Diarrhea in severely burned children. JPEN Journal of Parenteral and Enteral Nutrition. 2005;**29**:8-11

[111] Liolios A, Oropello JM, Benjamin E. Gastrointestinal complications in the intensive care unit. Clinics in Chest Medicine. 1999;**20**:329-345

[112] Wiesen P, Van Gossum A, Preiser JC. Diarrhoea in the critically ill. Current Opinion in Critical Care. 2006;**12**:149-154

[113] Labbe AC, Poirier L, Maccannell D, et al. *Clostridium difficile* infections in a Canadian tertiary care hospital before and during a regional epidemic associated with the BI/NAP1/027 strain. Antimicrobial Agents and Chemotherapy. 2008;**52**:3180-3187

[114] Karanika S, Paudel S, Zervou FN, Grigoras C, Zacharioudakis IM, Mylonakis E. Prevalence and clinical outcomes of *Clostridium difficile* infection in the intensive care unit: A systematic review and Meta-Analysis. Open Forum Infectious Diseases. 2015;**3**(1): ofv186

[115] Rubin MS, Bodenstein LE, Kent KC. Severe *Clostridium difficile* colitis. Diseases of the Colon and Rectum. 1995;**38**:350-354

[116] Grundfest-Broniatowski S, Quader M, Alexander F, et al. *Clostridium difficile* colitis in the critically ill. Diseases of the Colon and Rectum. 1996;**39**:619-623

[117] Ali SO, Welch JP, Dring RJ. Early surgical intervention for fulminant pseudomembranous colitis. The American Surgeon. 2008;**74**:20-26

[118] Byrn JC, Maun DC, Gingold DS, et al. Predictors of mortality after colectomy for fulminant *Clostridium difficile* colitis. Archives of Surgery. 2008;**143**:150-154

[119] Sebaihia M, Wren BW, Mullany P, Fairweather NF, Minton N, et al. The multi-drug resistant human pathogen *Clostridium difficile* has a highly mobile, mosaic genome. Nature Genetics. 2006;38:779-786

[120] Pichenot M, et al. Fidaxomicin for treatment of *Clostridium difficile* infection in clinical practice: A prospective cohort study in a French University Hospital. Infection. 2017 (Forthcoming).

[121] McFarland LV, Clarridge JE, Beneda HW, Raugi GJ. Fluoroquinolone use and risk factors for *Clostridium Difficile*-associated disease within a Veterans Administration health care system. Clinical Infectious Diseases. 2007;**45**(9):1141-1151

[122] Biller P, Shank B, Lind L, et al. Moxifloxacin therapy as a risk factor for *Clostridium difficile*-associated disease during an outbreak: attempts to control a new epidemic strain. Infection Control and Hospital Epidemiology. 2007;**28**(2):198-201

[123] Study of a Candidate *Clostridium Difficile* Toxoid Vaccine (Cdiffense) in Subjects at Risk for *C. Difficile* Infection [Internet]. 2016. Available from: https://clinicaltrials.gov/ct2/show/NCT01887912 [Accessed: March 2, 2017]